Oakland Fire Department

1869 - 2004

West Grand by Danny Barlogio

Turner®
PUBLISHING COMPANY

412 Broadway • P.O. Box 3101
Paducah, Kentucky 42002-3101
270-443-0121
www.turnerpublishing.com

Copyright © 2004 Turner Publishing
Company. All rights reserved.
Turner Publishing Company Staff:
Editor: Randy Baumgardner
Book Designer: Ina F. Morse

This book or any part thereof may not be
reproduced without the written consent of
the publisher.

Library of Congress Control No.
2003111778
ISBN: 1-56311-928-5

Printed in the United States of America.
Additional copies may be purchased directly
from the publisher. Limited edition.

Rick Chew

TABLE OF CONTENTS

Letter from Gerald A. Simon 4
Letter From Robert C. Bobb 5
Letter from Jerry Brown 6
History of Oakland Fire Department 9
OFD Organization - Present 38
Fire Chiefs of the Oakland Fire Department 39
Fire Safety & Education Day (Week) 43
Random Acts of Kindness 45
Office of Emergency (OES) 46
Oakland Black Firefighters Association 47
Line of Duty Deaths 49
Personnel 50
Fire Stations 84
Photo Album 94
Index 135

Getting to the peak.

CITY OF OAKLAND

IVE OFFICES • 150 FRANK H. OGAWA PLAZA • SUITE 3354 • OAKLAND, CALIFORNIA 94612

Department

(510) 238-3856
FAX (510) 238-7924
TDD (510) 238-3254

Greetings,

On behalf of the proud men and women of the Oakland Fire Department, it is my honor to invite you to share in the past, present and future of our journey in the professional fire service.

When we look into our history it is evident that Oakland is no stranger to disasters and emergencies of all types. The Oakland Firestorm of 1991, the Loma Prieta Earthquake of 1989, the Gasoline Tanker in Caldecott Tunnel Fire of 1982, BART Fire of January 1979, and the Army Base Fire of 1949, are some historical examples that illustrate this point.

Through it all, one thing remains constant: Oakland Firefighters have always been there, have always worked tirelessly and have always served with dignity and distinction.

Our Firefighters and all of the support staff of civilians and sworn personnel are truly the glue that holds it all together.

As we continue on our journey toward being a world class department, I invite you to share with us through this yearbook the rich tradition, the countless sacrifices and the special nature of our people who have honored, protected and served the citizens of Oakland since 1869.

We are courageous, committed and compassionate professionals who are destined, through utilizing our strengths and abilities, to become a world class fire department by the year 2010. It is with pride that I salute each of you who make us the great department we are.

Enjoy this journey through our history.

Sincerely,

Gerald A. Simon
Fire Chief

CITY OF OAKLAND

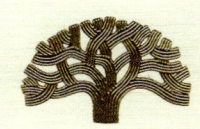

1 FRANK H. OGAWA PLAZA • OAKLAND, CALIFORNIA 94612

City Manager

(510) 238-3301
FAX: (510) 238-2223
TDD: (510) 238-2007

June 23, 2003

As City Manager, I would like to express my sincere gratitude for the service you provide to the City of Oakland. The vital role of Fire Fighters is respected not only by the city administration, but by the community as well. We thank you for the hard work and dedication that continues to save countless lives each year.

Sincerely,

Robert C. Bobb
City Manager

CITY OF OAKLAND

A PLAZA • 3RD FLOOR • OAKLAND, CALIFORNIA 94612

(510) 238-3141
FAX: (510) 238-4731
TDD: (510) 839-6451

IN RECOGNITION

It is my pleasure and honor to extend special recognition to the men and women of the Oakland Fire Department and thank them for their dedicated service to the people of Oakland.

Day in and day out, the members of the Oakland Fire Department respond selflessly to the call of duty, risking their lives to protect others. Oaklanders are fortunate to have people with your courage, strength and determination working to ensure that our city is safe and that we are prepared for whatever threats to our safety may arise.

Thank you for your commitment to the health and well-being of the people of Oakland.

Respectfully,

JERRY BROWN

Fire at California Spa, August 1999, High St. Exit and Coliseum, Four Alarms. Paul Fellows, Carlos Harvey and Carl Gardner, 18 E - 6 T. (Photo courtesy of Paul G. Fellows)

7

Station 8, 1901

History of the Oakland Fire Department

#1 Ladder truck.

Engine Co. 1, late 1800s.

THE HISTORY OF THE
OAKLAND FIRE DEPARTMENT

1853-1869 The Oakland Fire Department's Origins

The Oakland Fire Department had its beginnings in the volunteer fire companies that dotted the east shore of San Francisco Bay in the mid-1800s. The first attempt to organize a fire department in Oakland was made in 1853 with the formation of three companies (Empire Engine, Washington Engine, and Oakland Hook and Ladder) and the election of John W. Scott as Chief Engineer. These early companies were located in the vicinity of 5th and Washington Streets. After two years, the city council disbanded the companies for inefficiency, and their equipment and property were distributed to other volunteer companies, such as the Hancock Fire Brigade, a military organization. During this period, the city council attempted to reduce fire losses by restricting the fire response limits to lots only along Broadway, and prohibiting the erection of wood frame buildings in this area.

March 13, 1869 The OFD is Officially Organized

The present Oakland Fire Department was formed on March 13, 1869, by veterans under the direction of Col. John W. Scott, Oakland's first "volunteer" Chief Engineer. Equipment consisted of an Amoskeag steam pumper, a hose cart on loan from Col. Scott, a few feet of hose, and some minor appliances. Officers elected were J. C. Halley, Chief Engineer; Thomas McGuire, First Assistant; and George Taylor, Second Assistant. Phoenix Engine Company No. 1 was located on 15th Street in a one-story frame building believed to have been built for the department in 1854 and remodeled for the new organization. Washington Engine Company No. 2 and the Relief Hook and Ladder Company were located in temporary quarters until the completion of their new house at 5th and Washington Streets in 1870. This two-story frame building was ironically destroyed by fire; however, the apparatus and horses were saved by the valiant efforts of the firefighters. All early OFD records were destroyed in this blaze.

Fires of the 1860s

In November, 1862, Becht's Brewery at 9th and Broadway suffered a fire loss of $6,000 and 2,000 gallons of beer (the latter causing much distress among the firemen). On March 25, 1865, a disastrous fire occurred in the Delger block, situated at 14th and Broadway, and destroyed most of the block for a loss of $50,000. The city council was accused by the

Oakland News of being "penny wise and pound foolish" for not providing $1,000 worth of firefighting equipment that may have prevented the loss. Because of the renewed agitation for better fire protection, a meeting was called and the Oakland Guard, a military organization, was given permission to form a hook and ladder company.

1870s

The city purchased a new steam pumper, but could not afford to have it shipped from the east coast. Mayor Felton came to the rescue by paying the freight costs, and Engine 2 changed its name to J. B. Felton Steam Fire Engine, No. 2, in his honor. The quarters of Engine 2 would be referred to as the "Felton House" for years to come.

By 1872, the city had annexed territory as far north as 36th Street and east to about 23rd Avenue with the acquisition of Brooklyn, a large town just east of Lake Merritt. On April 8, 1872, the West Oakland Hose Company was admitted into the Oakland Fire Department to provide fire protection for the "Point," as West Oakland was then known. Henceforth, this company would be Engine 3 of the OFD. On December 11, 1872, Brooklyn Engine Company No. 3 was admitted to the department and became Oakland's Engine 4. In his report to the city council dated April 15, 1872, newly appointed Chief Engineer George Taylor lists his department's equipment as "one steam fire engine, one two-wheel hose carriage, one hook and ladder truck and hooks, 2,100 feet of fire hose, 41 fire hats and belts, and two fire bells." The department consisted of Phoenix Engine 1, with 45 members, and Relief Hook and Ladder Company, with 37 members. Chief Taylor also made many recommendations for improvements in his report, including an increase in water main sizes, additional cisterns, and the building of a second engine house in the lower portion of the city. On October 7, 1872, his hook and ladder company complement was increased to 70 men. On November 4, 1872, 35 citizens organized into the department in place of the disbanded Felton Engine Company.

During his two years in office (1872-1874), Fire Chief George Taylor guided the growth of the force, oversaw the installation of the fire alarm system started in 1873, and planned the first of many reorganizations that was finally approved by the city council in August, 1874.

By 1876, the City of Oakland had a population of 30,000. The OFD now consisted of three steam engine companies, a hose company, and a hook and ladder company. A new fire station had been approved for East Oakland (Engine 4) and another one had just been opened on 8th Street in West Oakland (Engine 3). A headquarters station had been completed in 1875 on 6th Street near Broadway (Engine 2, Truck 1, and Hose 1). After two years of construction, the $7,500 Police Telegraph and Fire Alarm System was completed with circuits for 11 fire alarm boxes terminating at the fire alarm station at 13th and Washington Streets. Oakland's firefighting force now had a total of 39 officers and men, 15 under full pay and 24 extra men. Eleven fire horses were "all in good condition and well trained to the business," Chief M. de la Montanya reported in the 1876 Annual Report. The report also indicated that salaries accounted for $13,314 out of a total budget of $28,903. The total

7 Engine, 1903.

value of fire department property and equipment was just over $60,000. Firefighters responded to 39 alarms of fire (26 fires, six false alarms, and seven out-of-city responses) during the year, with a total fire loss of $24,410. In 1877, the OFD accepted a new automatic repeater for the fire alarm system at a cost of $1,000.

Bids were being accepted for a Hayes Patent Fire Escape Truck, to be built to the specifications of Truck 2 of the San Francisco Fire Department. This new fire truck would go into service in Oakland as Relief Hook and Ladder 1. This new truck, with an 85 ft. wooden aerial ladder, was commended as being an "invaluable adjunct" by the Fire Chief, who added that he believed it had "no equal in the United States." He recommended that the old truck be repaired and placed in service in East Oakland, and suggested that with one horse and driver it would be of great assistance in that locality. This was accomplished in 1880, when Clinton Hook and Ladder, No. 2, was placed in service. Daniel D. Hayes, inventor of the first aerial ladder, worked as Superintendent of Steamers for the SFFD. He lived and had two workshops in Oakland, and it was here that these first aerial ladders were built.

Oakland's first professional Fire Chief was elected on January 7, 1878, approved by the city council, and appointed to the post on February 4, 1878. At the time of his appointment, James Hill was hose cart driver at Engine 2. His fire service career began with the Fire Department of New York in 1834. He later moved to Cleveland, Ohio, where he served as Chief Engineer of that department for 17 years. His expertise in fire department matters was the beginning of planned fire fighting activities in Oakland. He introduced discipline, and made the foremen responsible for the actions of the firefighters under their direction. Many changes were made to increase the efficiency of the department and a set of rules and regulations were introduced and enforced by Chief Hill, with the backing of the city council. He ordered all hydrants painted white for better visibility at night. He formalized assignment rules with the introduction of the "fifteen minute rule," which instructed companies to cover-in to certain locations when an alarm was struck. If no second alarm was struck during the fifteen minutes, the companies were to return to their own quarters.

In his 1878 Annual Report, Chief Hill stated that the department responded to 44 alarms of fire, with two false alarms. The budget for the year was $33,620, with salaries for the 45 men in the department accounting for $19,300. The report further stated that "all of the houses are in good substantial condition except that of the Phoenix Engine, which is in a dilapidated condition entirely unfit as to convenience and unhealthy for the men." The Phoenix Engine house was later reconditioned to make the building livable, and served the company until the building was replaced with a three-story brick station about 1885.

In February, 1879, special orders were issued to all companies stating that Broadway was not to be used responding or returning from alarms unless absolutely necessary "for the purpose of avoiding running over pedestrians, as much as possible." The "boys" had traveled too fast on Broadway once too often.

Fires of the 1870s

Two large fires occurred during Chief Fuller's eight months in office. The first, a fire in the city hall at 14th and Washington Streets, occurred on August 25, 1877, at 9:15 p.m. First arriving firefighters found the building well involved and quickly sent in a second alarm which brought the remaining two companies to the scene. The entire department could not save the building, which had been built in 1869 at a cost of $70,000. While the bell tower was being built next to the Phoenix Firehouse to replace the one lost in the city hall, fire alarms were sent from the bell in the Presbyterian Church.

The second fire occurred in the early morning hours of November 30, 1877, when the Oakland Guards Armory on 13th Street suffered a heavy loss. The Tribune commented that "they did some good work, and by their combined efforts kept the fire from spreading." The Armory was a tall, two-story frame building between larger brick buildings, which helped prevent the spread of the fire.

OFD Organization Circa 1875

Battalion 2	6th Street & Washington Street
Engine 1	15th Street & Washington Street
Engine 2	6th Street & Washington Street
Engine 3	8th Street & Willow Street
Engine 4	E. 14th Street & 12th Avenue
Hose 1	6th Street & Washington Street
Truck 1	6th Street & Washington Street

1880s

To protect the rapidly growing northern section of the city, Hose 2 was placed in service on February 22, 1880, at San Pablo Road and Market Street. Hose 2 had a four-wheel hose reel with 1100 feet of hose. Contra Costa Water Company's mains in the area were more than adequate, with the supply from Lake Temescal making the hose streams effective. The company remained until it was replaced by Engine Company 5, at Milton and Market Streets, in 1885.

In the 1880s, the Chief Engineer, who received $150 per month, was responsible for the administration of routine and firefighting operations. The First Assistant Engineer was the master mechanic who tended to the repairs of apparatus and equipment, primarily the steam fire engines. He assumed the Chief Engineer's duties during an absence. The Second Assistant Engineer was third in command and had the day-to-day responsibilities for all hose and hydrants, assisted by the engineers in the fire companies if necessary. Fire Warden George H. Carlton was responsible for the fire alarm system, which consisted of both the Police Telegraph boxes and 46 fire alarm boxes.

Very little is known about the administration of Chief Engineer James Moffitt (1883-1889). He had been associated with both the San Francisco and Oakland Fire Departments, was a founder of the Brooklyn Volunteers (Engine 4, Truck 2), and was appointed Chief Officer sometime after 1878. He was appointed Chief Engineer in 1883, and served until his death in April, 1889, at age 58.

James F. Kennedy was appointed Chief Engineer on April 23, 1889. Kennedy was born in Ohio, moved to Iowa, and began his career as a firefighter there at the age of 16. He was a Foreman (Captain) before he was 21 and was later made Assistant Chief, then Chief Engineer, of the Keokuk (Iowa) Fire Department. He moved to California in 1876, eventually settling in Oakland. He served as an assistant under Chief Moffitt and became Chief Engineer when Moffitt died. Chief Kennedy retired in 1893, and later went to work for the federal government at the Mint until 1919.

By 1885, the Oakland Fire Department consisted of four engine companies (Phoenix No. 1, in a new three-story brick headquarters on 15th Street behind city hall; Felton No. 2, on 6th Street between Broadway and Washington Street; Point No. 3, on 8th Street between Willow and Wood Streets; and Brooklyn No. 4 on 14th Street between 12th and 13th Avenues), two hose companies (Hose 1 with Engine 2, and Hose 2 on San Pablo at Market Street), and two hook and ladder companies (Relief No.1, with the Felton En-

Auditorium raise at 3 Engine.

The truck ladders old, old 2 Engine.

gine, and Clinton No. 2, with the Brooklyn Engine Company). A new two-story house for Engine 5 was under construction at the corner of Milton and Market Streets to replace Hose 2. This addition would give the city's northern section better fire protection, and increase the Oakland Fire Department's pumping capacity by 20%.

Fires of the 1880s

By 1880, the city had a population of 34,555 and was the site of many fine homes and commercial buildings. One of Oakland's finest hotels was the Grand Central, located on 12th Street between Webster and Harrison, built by Dr. Samuel Merritt and others in 1873. The fire began at 1:00 a.m. on March 9, 1880, during a northwest gale, but was not discovered until 2:00 a.m. The Weber House and four one-story buildings on the same block were also destroyed.

Another spectacular fire occurred in the early morning hours of September 9, 1880, at the Galindo Hotel on 8th Street. The building suffered a $50,000 loss. A repeat of this fire occurred 91 years later on November 17, 1971, and went to five alarms before being brought under control. This was the last fire for the Galindo Hotel, which was declared a total loss.

In an editorial dated Saturday April 29, 1882, the Oakland press commended Chief Hill for his department's actions at the Oakland Planing Mill Fire. Companies stopped the spread of fire to eight exposures, one of which was Freeman and Smith's Coal Yard. A fire with a $10,000 loss occurred at the Pacific Jute Mill, 6th Avenue and E. 12th Street, on May 29, 1882. It took half the department all night to overhaul the jute bales.

OFD Organization Circa 1885

Battalion 2	15th Street & Washington Street
Engine 1	15th Street & Washington Street
Engine 2	6th Street & Washington Street
Engine 3	8th Street & Willow Street
Engine 4	E. 14th Street & 12th Avenue
Hose 1	6th Street & Washington Street
Hose 2	San Pablo Avenue & Market Street
Truck 1	6th Street & Washington Street
Truck 2	E. 14th Street & 12th Avenue

1890s

In 1890, with a population approaching 50,000, the Oakland Fire Department began to feel the positive effects of the city's growth. The Fire Commission authorized the addition of one foreman and four extramen each for Truck 2 in East Oakland and Hose 2 in West Oakland.

Chemical 1, a buckboard with two 40-gallon chemical (soda-acid) tanks, was placed in service on Webster near 13th Street. Chemical 1 would serve downtown Oakland from this location until 1910. Chemical 2 was located on Magnolia just below 34th Street in a two-story frame station. Chemical 2 would serve North Oakland until 1924, when the station and the company were replaced by Engine Company 22. Chemical 3 was put in service in 1891, on East 14th Street and 17th Avenue, in a one-story building. In 1903, Chemical 3 relocated to Engine 6 and was disbanded on June 1, 1906.

In 1894, three companies were added to the roll of the department. A third ladder company (Truck 3) was placed in service in West Oakland on 8th Street next to Engine 3, in a one-story frame building. In 1896, these companies moved into a large two-story brick fire station on the same site. These two companies remained in that location until 1950, when a new fire station was built at 7th and Pine Streets. To fill the gap between Engines 2 and 3 in the fast growing west end, Hose 3 was placed in service at 7th Street near Union. In East Oakland, Engine 6, located at E. 15th Street between 22nd and 23rd Avenues and staffed with three permanent men, a foreman and five extramen, filled out the fire protection in the east end. In 1920, this building became so structurally weak that the companies moved out, and the building was replaced with a modern two story fire station.

Monthly salaries in 1894 ranged from $150 for the Chief Engineer, $125 for Assistant Chiefs, $75 for drivers and $20 for extramen. Extramen were generally assigned to companies near their place of business or home, so were almost always in a position to 'catch the rig' if the fire bell or whistle sounded. The signals were coded by box number (one, two and three digits) and each extraman carried a list of box numbers and locations with him. The use of extramen was gradually reduced as they were replaced by permanent (paid) men. Early Oakland Fire Department records show that extramen performed good and faithful service and could be depended upon to carry out firefighting operations in an efficient manner. Extramen were used until 1917, when the position was abolished by a city ordinance.

In 1897, the city annexed the remaining territory between Oakland and Berkeley, which added 20,000 citizens to the population and consolidated the districts of Golden Gate, Temescal, Alden, Linda Vista, Peralta and a portion of Piedmont. The citizens in the newly annexed area had little fire protection except for a group of young volunteers with a hose company known as the Temescal Hose Company. Their hose cart was later given to the Children's Home Society after Engine 7 was placed in service on San Pablo Road at about 58th Street. This temporary station served as Engine 7's quarters until a new one was completed in 1909, on 59th Street just off San Pablo Avenue.

Oakland's thirteenth Chief Engineer was appointed March 29, 1898. Nicholas A. Ball joined the department in 1880, and served as extraman with Engine 1. Ball was promoted to Foreman of Engine 1 in 1884, but a change in city administration caused him to resign in 1887. After the death of James Moffitt, the new Chief, James Kennedy, asked Ball to join his administration as First Assistant. He remained in this position until 1893, when the Board of Police and Fire Commissioners removed him. When he returned as Chief Engineer, the Oakland Fire Department consisted of over 100 men, six engine, three hose, three truck and three chemical companies, housed in ten fire stations. Fire Chief Ball would serve the City for over 17 years, becoming the first Fire Chief to overcome the political pressure of the City Council, which had always changed Fire Chiefs with each new administration at City Hall. Before his retirement in 1915, Chief Ball would see the City, and the Oakland Fire Department, double in size.

Fires of the 1890s

One of the most spectacular fires in the history of Oakland destroyed the Tubbs Hotel on the evening of August 14, 1893. Located at 5th Avenue and East 12th Street, the hotel was built in 1879 at a cost of $200,000. At the time of the fire, it was the largest building in Oakland. The Tubbs fire occurred on a clear summer night and was described by on-lookers as "the prettiest fire scene ever witnessed in the city." Unfortunately, the hotel was only insured for $25,000.

In 1885, St. Mary's College moved from San Francisco into a large brick building on Broadway at Hawthorne. A fire at St. Mary's, on September 23, 1894, started in a pile of rubbish in a waste chute, and burned for some time before being discovered. Despite the heroic efforts of firefighters, the fire burned

Engine and Truck at 4 Engine.

for three hours before being brought under control. The dollar loss for this fire was in excess of $100,000.

On the evening of November 24, 1897, a fire occurred at Sacred Heart Church, 40th and Grove Streets. Responding companies were Engines 1 and 5, Truck 1, Chemical 2, and Hose 3. The long response time for this fire pointed out the need for better fire protection in North Oakland.

Chief Ball would prove his ability to handle large fires many times during his career. His first second alarm as Chief occurred on July 5, 1898, on 9th and Webster Streets, for a fire in a blacksmith's shop. His first general alarm fire, calling out the entire department, was for a fire in the middle of the block on 28th Street near West on the night of February 22, 1899.

1900s

By 1900, Oakland had a population of 67,000 and had become a major rail terminus and seaport. With the recent annexation of all territory north to the Berkeley line, citizens began to pressure the city to live up to its promise to provide fire protection in this area.

Using apparatus and equipment purchased the year before, Chief Ball had temporary quarters built in both the Alden and Golden Gate districts. On July 1, 1901, Engine 7, on San Pablo Road at 58th Street, and Engine 8, on 51st Street just east of Telegraph Avenue, were placed in service. The addition of these companies not only provided much needed services to North Oakland but increased the ability of the department to handle large fires and still have some reserve force available. These frame buildings would be used for eight years before being replaced with permanent structures.

In Linda Vista, a citizens' group was collecting money for a new fire house. Of the $7,000 collected for the project, $2,000 went for the plans and $5,000 for the lot at 165 Santa Clara Avenue. The city put up $5,000 for the building and furnishings, and, on August 20, 1904, Truck Company No. 4 was placed in service, providing ladder coverage for all of Oakland above 20th Street to the Berkeley line.

Chief Ball had another fight on his hands in 1905, after the National Board of Fire Underwriters published its report on the City of Oakland. This was the first of many surveys that the Underwriters would conduct to measure the fire defenses of the community. Ball's critics felt that the recommendations made in the report proved his inability to command the department. In his defense, he showed the city fathers seven years of annual reports outlining the same problems the National Board had found and his recommendations to correct these deficiencies. On the positive side, the public airing made the average citizen aware of the fire protection problems in the city.

Old 5 Engine.

Fruitvale Company #1.

The OFD continued to expand during the next two years. A new fire station on 59th Street was built for Engine 7 and Truck 5 in 1907, providing much needed ladder coverage for North Oakland. In 1908, Truck 4 was relocated to rented quarters at 26th and Broadway, awaiting the completion of their new station on 25th Street. Engine 10 was placed in service at the Santa Clara venue station. Later in the year, Engine 8's new firehouse on 51st Street was completed, and served as their quarters until the early 1950s. In 1909, Hose 3, at 7th and Filbert, was replaced by Engine 9 on Magnolia below 14th Street.

The population of the city doubled between 1900 and 1910. The city now claimed 150,174 citizens.

The largest factors in this growth were the 1906 earthquake, which added about 65,000 permanent residents, and the annexation of county land to the City in 1909, adding the districts of Claremont, Fruitvale, Melrose, Fitchburg and Elmhurst to the city. Total area now belonging to the City of Oakland was approximately 60 sq. miles. The City added a few paid men, but continued to use the former volunteers as extramen.

Engine 11 was placed in service in 1909 on Alice Street, replacing Chemical 1 at 13th and Webster. Engine 11's former quarters still stand today, in use as a private residence.

Fires of the 1900s

One of Oakland's largest fires took place on August 10, 1902, when the First Baptist Church at 14th and Brush was destroyed. The first alarm was sent in at 10:06 p.m., quickly followed by the second at 10:12 p.m. and the third (general) alarm at 10:29 p.m. on Box 54, 12th and Brush Streets. This large, wood building had been built in 1868, at a cost of $35,000, and was a landmark in Oakland. Using the entire department, the strategy employed by Chief Ball was to mass large hand lines between the fire and the exposures. The plan worked, and the block was saved.

The Oakland Fire Department's first line-of-duty death occurred on July 5, 1903. The fire involved the Arcata Lodging House in the Pardee block on San Pablo just below 14th Street. Frank Parker, an extra man with Engine 3, was killed when the roof and walls collapsed on the crew. Foreman A. A. Sicotia and Fireman Daniel J. Barr were seriously injured. Many citizens donated money for relief for the Parker family and the injured men.

A fire that started under the Model Lodging House at 8th and Webster Streets caused the department some concern before being brought under control. The Evening Tribune described the place as "one of the worst fire traps in the city," indicating that the owner, a local doctor, had been cited for gross violations of the fire safety laws when the buildings were being built. The entire department was required to keep this loss to one-half of the block and $50,000.

In April, 1906, mutual aid was sent to San Francisco via barges to assist in combating the conflagration following the earthquake. Several Oakland engine companies successfully assisted the SFFD in their efforts to save portions of the City.

OFD Organization Circa 1901

Battalion 2	15th Street & Washington Street
Engine 1	15th Street & Washington Street
Engine 2	6th Street & Washington Street
Engine 3	8th Street & Willow Street
Engine 4	E. 14th Street & 12th Avenue
Engine 5	Milton Street & Market Street
Engine 6	E. 15th Street & Munson Street
Engine 7	58th Street & San Pablo Avenue
Engine 8	51st Street & Telegraph Avenue
Hose 1	6th Street & Washington Street
Hose 2	San Pablo Avenue & Market Street
Hose 3	7th Street & Union Street
Chemical 1	13th Street & Webster Street
Chemical 2	34th Street & Magnolia Street
Chemical 3	E. 14th Street & 17th Avenue
Truck 1	6th Street & Washington Street
Truck 2	E. 14th Street & 12th Avenue
Truck 3	8th Street & Willow Street

Station 8 at 466 51st Street.

1910s

The first steps to eliminate the position of extraman were taken in July, 1911, when Engines 1, 2, 11 and Truck 1 were made permanent, fully-paid companies. The men for these positions came from the ranks of the extramen where possible. Engine 19, at 56th and Dover, became the next fully-paid company using a new motor-driven hose and chemical wagon. During fiscal year 1911-12, the department added eighty-four men, new steamers, hose wagons with deck guns, and chiefs' autos. A lot on Allendale and Abbey was purchased for Engine 17 and the E.B. and A.L. Stone Company donated land to the city at 93rd Avenue and E. 14th Street for a new fire station to house Engine 20.

W.B. Smith

In 1912, the department added more men, equipment and stations. On College Avenue, Chemical 1 was formed. On 25th Street near Broadway, Engine 15 was established and was joined by Truck 4. On 13th Avenue at Hopkins Street, the OFD placed Engine 16. To make way for the New City Hall, the old headquarters fire station was torn down, and the fire alarm equipment was relocated to the new central fire alarm station on Oak Street. Engines 1 and 12 were relocated to 18th Street just east of San Pablo. When completed in 1914, the new City Hall had space for the Chief's offices, and Engines 1 and 12.

When Chief Ball retired in 1915, the Oakland Fire Department consisted of three hundred and three men in twenty-two stations. Battalion Chief Elliott Whitehead succeeded Ball as Chief Engineer. Many changes took place during the next six years. Chief Whitehead directed the removal of the extramen (placing the Oakland Fire Department on a fully-paid basis) in 1917, and established a Fire Protection Bureau the following year. He also made many other improvements recommended by the 1910 National Board of Fire Underwriters survey of the City of Oakland. Whitehead's plan for the complete motorization of the department was accomplished in 1922. During his tenure as Fire Chief, he saved taxpayers a large sum of money by building apparatus and equipment in the city shops, under the direction of Shop Superintendent James M. Ready and Fire Department Superintendent of Engines J. E. McFeeley.

Fires of the 1910s

Oakland had many serious fires during World War I along the waterfront in the shipyards. Another major fire occurred in August, 1918, when the Metropole Hotel at 13th and Jefferson Streets was destroyed.

OFD Organization Circa 1914

Battalion 2	15th Street & Washington Street
Battalion 3	E. 15th Street & Munson Street
Battalion 4	Champion Street & Palmetto Street
Battalion 5	25th Street & Telegraph Avenue
Battalion 6	51st Street & Telegraph Avenue
Engine 1	15th Street & Washington Street
Engine 2	6th Street & Washington Street
Engine 3	8th Street & Willow Street
Engine 4	E. 14th Street & 12th Avenue
Engine 5	Milton Street & Market Street
Engine 6	E. 15th Street & Munson Street
Engine 7	58th Street & San Pablo Avenue
Engine 8	51st Street & Telegraph Avenue
Engine 9	14th Street & Magnolia Street
Engine 10	Santa Clara Street & Vernon Street
Engine 11	14th Street & Oak Street
Engine 12	14th Street & Washington Street
Engine 13	E. 12th Street & 33rd Avenue
Engine 14	Champion Street & Palmetto Street
Engine 15	25th Street & Telegraph Avenue
Engine 16	13th Avenue & Hopkins Street
Engine 17	Allendale Avenue & Abbey Street
Hose Chemical 1	E. 14th Street & 50th Avenue Became Engine 18
Hose Chemical 2	56th Street & Dover Street Became Engine 19
Hose Chemical 3	E. 14th Street & 93rd Avenue Became Engine 20
Hose Chemical 4	42nd Street & Montgomery Street Became Engine 21
Chemical 1	College Avenue & Birch Court Became Engine 32
Chemical 2	34th Street & Magnolia Street Became Engine 22
Truck 1	6th Street & Washington Street
Truck 2	E. 14th Street & 12th Avenue
Truck 3	8th Street & Willow Street
Truck 4	25th Street & Telegraph Avenue
Truck 5	59th Street & San Pablo Avenue

1920s - The End of the OFD's Horse-Drawn Era

One of the most interesting phases of Oakland Fire Department History was the complete motorization of the department in the early 1920s, the end of the "horse-drawn era." Horses had been faithfully pulling a variety of fire apparatus (steamers, hose carts, ladder trucks, and chiefs' buggies) to fires in Oakland since the late 1860s. The passing of the gallant fire horses drawing the picturesque steam engines as they raced smoking down the streets of Oakland was not accomplished overnight. Due to many problems encountered with early motorized apparatus, the transition from horses to modern engines took almost a decade. Departmental records are vague, but it appears that the last horse-drawn companies remained in active service in Oakland until early 1921.

The life of a fire horse in Oakland was hard, even though the horses were accorded the best of care given to any animal of the time. Horses were confined to stalls located next to their apparatus for most of the day. Departmental rules required that the horses be exercised each day, weather permitting. This was usually done between the hours of 1:00 p.m. and 3:00 p.m. on the streets near quarters with apparatus and crew. Firefighters fed and cared for the horses as part of their daily duties. Stalls were equipped with an electrically operated chain that would drop automatically when an alarm was received. The horses were trained to be lead out and stand beneath their harnesses, which were suspended from the ceiling. Once the horse was in position under the combination harness and open swivel collar, the driver would release the mechanism and the harness would drop over the horse. This system was known as the "lightning hitch," as it was an extremely efficient method of hitching up a team of fire horses in a very short amount of time. Once the harness had dropped, a collar was snapped around the neck of the horse by firefighters, and the apparatus was on its way in a very short time.

One popular myth about fire horses that has survived is that the horses were made to gallop when responding to fires. Fire horses were only made to gallop when photographers wished to snap a picture of these brave animals "in action." In reality, the fire horses normally pulled their heavy apparatus at a fast walk, and upon reaching the scene of a working fire, were unhitched, blanketed, and taken to a safe place away from the noise and sparks given off by the working steamers.

11 or 12 Engine

Pump test of new Engines, 1920s.

The main problem relative to final motorization of the department was the concern of Fire Chiefs, not only here in Oakland, but across the nation, that the small size and lack of dependability of the early-motorized fire engines would render them ineffective. Horses never failed to start, as these early engines often did. Another strike against motorized fire engines was their low-capacity, motor-driven pumps. The horse-drawn steam pumper of the 1920s had reached a state of high efficiency. For example, by the turn-of-the-century, the Fire Department of New York had horse-drawn steam engines capable of pumping over 1500 GPM. As late as 1912, the FDNY continued to purchase new horse-drawn steam engines, rather than buy untried, under-powered motor apparatus.

Oakland's first steps toward motorization were taken when the department began replacing the small buggies drawn by a single horse that were used by chief officers. The first chief's car was placed into service by the OFD in 1908. These new "buggies" were a variety of Columbia, Velies, Lozier, and Cadillac automobiles. Some motor-driven hose and chemical wagons were purchased in 1912 by the OFD from the Seagrave Company. Engine 12 and Engine 19 (as they were the first and last apparatus painted white used by the OFD) used these "White Wagons." These new apparatus were chemical wagons, and were used in place of regular pumping engines.

The first specially-built motorized fire engine purchased by the Oakland Fire Department was a Gorham, a mammoth machine that looked like a giant roadster. This Gorham Turbine Pumping Engine was built in Alameda and had a newly developed pump that was later the basis for the now famous Seagrave centrifugal pumpers used by many modern fire departments. The Gorham was purchased by the OFD in 1913, along with a motorized Webb Hose Wagon, and was put into service as Engine 1. Engine 1 and Engine 12 were in temporary quarters together on 18th Street awaiting the completion of their new quarters at City Hall.

Eng. Co. No. 18, Feb. 18, 1925.

Truck Co. No. 6, Feb. 18, 1925.

Engine 11's horse-drawn wagon and steamer were retired and replaced with a high-pressure wagon. This company was renumbered Engine 12. The 1000 GPM Gorham pumper originally at Engine 1 was transferred to Engine 12. The designation for Engine 11 was given to the High-Pressure Pumping Plant. Staffed by an officer and an engineer, Engine 11 was used to boost pressure to augment the downtown water supply in case of natural or man-made disaster. This old OFD building is still located on Oak Street and 14th Street and is currently used by the Oakland Park and Recreation Department for offices.

Early attempts to motorize the OFD truck companies were crude by today's standards. In 1914, small 4-cylinder Cadillac Roadsters, or touring cars, were utilized as tractors hauling old horse-drawn ladder trucks, but this proved unworkable. The "modern" tractors just didn't have as much "horse power" as a team of Oakland fire horses. In 1915, the OFD bought a fully motorized Seagrave "side saddle" aerial ladder truck. This unique contraption had an 85-foot wooden aerial ladder with a turntable located ahead of the driver's seat. In 1917, the OFD purchased another motorized Seagrave ladder truck. These two trucks went into service as Truck 1 and Truck 2, respectively.

The last horse-drawn companies in the OFD were Engine 2, on 6th Street between Washington and Broadway; Engine 3 on 8th Street between Willow and Wood Streets; Engine 4 at 1235 E. 14th Street; Engine 5 at Milton and Market Streets; Engine 7 on 59th Street just east of San Pablo; and Engine 11 on Alice Street between 8th and 9th Streets. The glorious days of the fire horse came to an abrupt end when the remainder of Oakland's companies received newer motorized apparatus. It is thought that Engine 5 was the OFD's very last company to be motorized.

1920s, Continued

On July 1, 1921, Samuel H. Short was appointed Chief of the Fire Department. Chief Whitehead returned to the rank of Battalion Chief to complete his 25 years. He later would have to go to court to secure his pension earned as Chief of the Department, a fight he won. Chief Short continued the improvements started by Chief Whitehead and had the OFD fully-motorized in 1922.

At the close of WWI, the Oakland Fire Department would continue its evolution from a mostly volunteer organization to a fully-paid, professional mu-

Eng. Co. 18, Truck Co. No. 6, Feb. 18, 1925.

Truck Co. No. 6, Eng. Co. No. 18, Feb. 18, 1925.

Seagrave

on January 1, 1920, adding 104 men to the department. The first African-American firefighters were hired by the OFD in 1922. These men were initially assigned to Engine 11, the High-Pressure Pumping Plant on the shores of Lake Merritt. During the next four years, many more African-Americans came to work for the OFD. The administration decided to move them all to newly established Engine 22, located at 34th Street and Magnolia Street in West Oakland. Thus began a 30-year period of racial segregation in the Oakland Fire Department.

In 1922, six new American-LaFrance pumpers were delivered to Engines 4, 6, 8, 9, 10 and 15, bringing the horse-drawn era to an end. During the next five years, 11 engines and two Seagrave trucks were delivered to replace most of the early-motorized equipment, much of it built in the city shops.

Plans were made to erect a new station for Engine 18, still located on 50th Avenue at E. 14th Street in the original Melrose Fire Department station. The new two-story station opened on December 1, 1924, with Engine 18, newly organized Truck 6, and Chief 4, formerly located at Engine 14. The addition of a truck company in East Oakland made the very long runs for Truck 2 no longer necessary. The Chiefs' district was redrawn, and Chief 3 was created and placed in service with Engine 6. East Oakland had previously been covered by an Assistant Chief and Battalion Chief 4 on alternate shifts.

The site for Engine Company 23 was donated by the Chevrolet Motor Company at 73rd Avenue and Foothill Blvd. New modern quarters were built to replace older stations for Engine 5 at Milton and Market Streets, Engine 13 on 33rd Avenue below E. 15th Street, and for Engine 22, newly organized to replace Chemical Company No. 2, at 34th and Magnolia Streets. Engine 22 was manned by black firefighters until 1953.

The last company to be added to the department for the next 22 years was Engine Company 24 on Moraga Road. The uniquely designed station served the Montclair district until 1993, when the Loma Prieta Earthquake made the building seismically unsafe for public use. Old Station 24 was commonly referred to by hill area residents as the "Montclair" Fire Department.

On July 11, 1927, William G. Lutkey was appointed Fire Chief. Chief Lutkey started his career in 1901 as driver of Chemical Co. 1, and was promoted to Foreman (Captain) of Engine 11 in 1909, Battal-

nicipal fire department. Steps were taken in early 1920 by department welfare organizations (such as the Health Club, Relief Fund and Fire Fighters Union - IAFF Local #55) to prepare certain charter amendments, including the adoption of a two-platoon system to further efficiency and strengthen civil service regulations as they applied to the OFD. These measures were approved by voters and went into effect

ion Chief in 1913, and to Assistant Chief in 1921. Lutkey's career as Chief of Department spanned three decades, during which the City of Oakland and its fire department continued to grow in size and stature.

Fires of the 1920s

During 1924, the city had a very high fire loss. In addition to a $400,000 loss at the Union Construction Company fire, 68 fires were traced to one youth, bringing the total loss for the year to $1,190,306.

1930s

The depression era was a no-growth period for the department. Chief Lutkey was faced with running the department without any increase in budget or personnel. The firefighters chose to give up salary to prevent layoffs of personnel.

Few changes took place during the 1930s. Engine 10's quarters were moved across Santa Clara Avenue in 1936 as part of a Works Progress Administration project. The frame building was remodeled, and faced with stucco and flagstone. Two new trucks were purchased in the late 1930s, going to Truck 2 and Truck 4.

Fires of the 1930s

The department still had to contend with the fire problems of a large city. In January 1930, the Fremont High School fire went to six alarms before being controlled. Another large and difficult fire to fight was the Key Route pier fire in 1933. This fire went to six alarms, plus special calls to San Francisco for fire boats. Engine 22 was placed on a special flat car and taken out on the pier to draft water for the fire. Overhaul continued for a number of days following the fire. In December, 1933, the Redwood Mill Company fire, near 57th and Los Angeles, went to six alarms before it could be stopped.

1940s

By 1940, the population of the city had reached 300,000. During World War II, this would increase to 450,000 people. The department was faced with many problems, such as losing 54 men to the various branches of service, hiring emergency men to fill those vacancies, and maintaining apparatus and equipment. During World War II, the OFD introduced fog nozzles and applicators (developed by Lloyd Layman for the U.S. Navy) to fight oil fires. The OFD also made provisions for fighting fires in special risk areas such as airplane hangers and large military warehouses by adding deck guns and other heavy stream appliances to their inventory.

The first reduction in working hours for firefighters came in 1946, when the 84-hour week was reduced to 72 hours. This reduction in hours required the addition of 118 men, bringing the total strength of the department to 502.

A milestone for diversity in the OFD occurred in the 1940s, when Royal Townes, a member of Engine 22 became the first African-American to be promoted from the ranks. Townes was appointed Chief's Operator and drove Chief 5 in downtown Oakland,

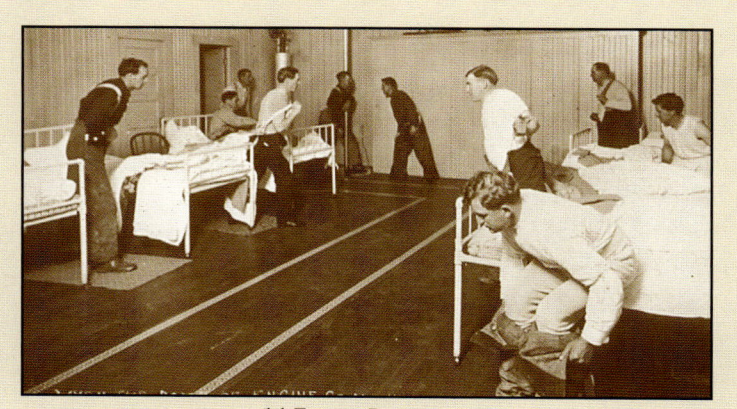

14 Engine Dormitory.

Waiting for the next Still at 14 Engine.

Superintendent of Engines and his crew.

10 Engine in original location.

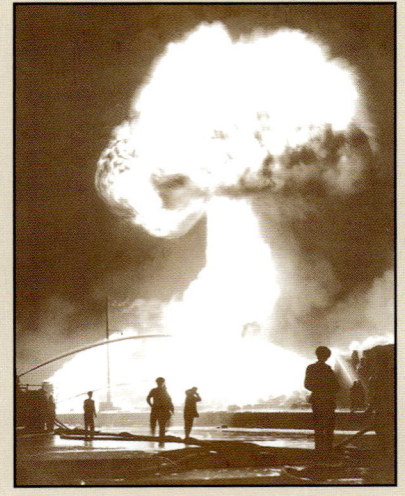

Army Base, 1943.

1948, 11 Engine, changed to 12 Engine.

4 Chief at 18 Engine.

Chief's Operator and Captain.

later promoting to the rank of Lieutenant at Engine 22. Several more African-American officers soon followed in Townes' footsteps, and by the latter part of the decade, Engine 22 became entirely segregated, with African-American officers and firefighters. Due to overcrowding at Engine 22, several more segregated firehouses were opened in Oakland during the late 1940s. These companies were Engine 28 on Skyline Blvd. near Roberts Park, and Engine 33 at Skyline Blvd. and Grizzly Peak Blvd.

A rehabilitation and improvement program was implemented in 1948 at a cost of over $1,275,000. Thirteen new fire stations were built; six older stations dating back to the late 1800s were closed. Deliveries of new apparatus and equipment long overdue, some due to war shortages, were made. Apparatus was purchased from Mack, Maxim, Pirsch and Coast to replace engines and trucks purchased in the 1920s. The 7th Battalion was added to give chief coverage to the fastest growing section of the city. The new companies added were Engines 25, 26, 27, 28, 29, 31, 33 and Trucks 8 and 9. Engine 30 was a fire boat tender to support the land operations of the new 10,000 GPM fireboat "Port of Oakland." Chemical Company 1 at City Hall was designated as Engine 32. Also included in the improvement program was a drill tower, located at 48th Avenue and MacArthur Blvd., built at a cost at over $24,000.

EMS 1940s style.

Old 16 Engine

USS *Hoga* YT-146/Port of Oakland

In 1948, the City and Port of Oakland were able to acquire, for the meager sum of $1.00 per year, a slightly worn U.S. Navy Tugboat. This vessel, the former USS *Hoga* (YT-146) would protect Oakland's waterfront for almost forty-five years, serving as the OFD's first fireboat. Prior to being leased by the Port and City of Oakland, the *Hoga* had served the Navy as a yard tug at Pearl Harbor Naval Station. She was there on that fateful Sunday of December 7, 1941, when the Imperial Japanese Navy carried out its devastating surprise attack on the unsuspecting U.S. Fleet. The USS *Hoga* operated non-stop from that Sunday until the following Wednesday, fighting fires on numerous ships and rescuing many sailors from Pearl's oily waters. She was able to save the USS *Oglala* from certain destruction by pushing her away from the doomed USS *Arizona*. But her greatest feat came late in the attack. The only battleship to get underway during the battle was the USS *Nevada*. The *Nevada*, though heavily damaged from several bomb and torpedo hits, was barely making way and attempting to escape the onslaught. She was heading for the open sea, but became the focus of the attacking Japanese aircraft. If the Japanese could sink the *Nevada* while she was in the narrow channel leading to the sea, she would become a "cork in a bottle," making it impossible for any other ship to get into or out of Pearl Harbor. Had the Japanese succeeded in sinking the USS *Nevada*, the outcome of the Pacific war could have been much different. Thankfully, the *Hoga* was able to come to the *Nevada*'s aid and ground her before the *Nevada* could sink in the channel. For her extraordinary work on that day of infamy, the USS *Hoga* and her crew received a special unit citation from Admiral Chester Nimitz.

In February, 1949, the new fire department building, a former supermarket and, later, ice rink, was commissioned at 14th and Grove Streets. This building, covering an entire city block, contained the administration, Fire Prevention Bureau, and personnel offices, as well as the fire department shop and supply section. In addition,

Old 5 Truck.

Old 13 Engine.

25

Old 14 Engine

Old 15 Engine

Old 22 Engine

Old Station 20.

1, 1955, when Fire Marshal James J. Sweeney, Jr. was appointed Fire Chief.

Fires of the 1940s

Fires of major importance during the late 1940s included the Fruitvale Canning Company, in September, 1948; West Coast Macaroni, in December, 1948; and one of the most remembered at the Oakland Army Base warehouse in March, 1949. This fire was the largest dollar loss in Oakland's history, at approximately $10 million dollars. The Oakland Fire Department had fought sixteen six-alarm fires during the last ten years, including the ones listed.

March 1949.

Old 7 Engine.

it housed Engines 1 and 32, Truck 1, a light wagon, a fuel supply wagon and extra chiefs' cars.

James H. Burke, who had risen through the ranks to become Chief of the Department, succeeded William Lutkey. He was faced with transforming an organization little changed since the mid-1920s. Chief Burke continued to purchase engine and truck apparatus in an orderly manner to replace older outdated equipment, increasing the efficiency of the department. A variation of this program is still in effect today. Chief Burke ended his 43 year career on July

Old Station 23 front.

31 Engine, 9 Truck.

Old 27 Engine

1950s

The department began another transition in the early fifties. The "atomic age" presented new problems for the fire service, and Chief Burke wanted to be prepared for any type of disaster that might threaten the city. Chief Burke built up the Civil Defense Auxiliary Fire Department with up to ten companies to assist in all types of emergencies.

Chief Sweeney was a young, educated, and progressive fire officer, and his experience as Fire Marshal would have much to do with the direction the department would take in the next ten years. Many changes took place within the organization with new emphasis placed on education and training programs. Special courses were given for firefighters and company officers in many subjects.

Soon after his appointment, Chief Sweeney began to plan for the future of the department. He is credited with the racial integration of the fire department by assigning black firefighters to all districts of the city. He continued to replace apparatus using the program started during the Burke administration. A plan was developed for the reorganization of the department. This plan included building new stations, relocating companies, and reducing personnel in some companies. The department would be able to begin working the 52-hour week when these changes were made. The three-platoon system went into effect on July 1, 1961. Chief Sweeney retired on January 1, 1972, after 38 years of service to the city.

Fires of the 1950s

The Oakland Fire Department gained national fame when ninety-one people were rescued with aerial and ground ladders in the early morning hours of September 24, 1951. The fire at the Oakland Court Apartments started near the first floor stairway, and quickly spread up through the upper floors of the building. Engine 1, returning from another alarm, passed the building just as the fire was discovered and made a direct attack on the fire by taking a big line up

29 Engine

Station 13

Station 23

Station 18

the stairway. Arriving companies were faced with people awaiting rescue at every window at this five-alarm fire. Hoseman Barney Walsh was awarded "Man of the Year" for his rescue of a man who jumped into his arms as he was reaching the top of Truck 4's aerial. The only fatality was a woman who failed to follow instructions from firefighters when she opened her door leading to the hallway and was fatally burned. The building still stands today, and is little changed since 1951.

Chief Sweeney's first greater alarm fire occurred just four days after his appointment, at the Canned Foods Distribution Center, a supermarket on E. 12th Street. This fire, with a loss of nearly $500,000, went to four alarms before being brought under control. While this fire was being mopped up another fire occurred on E. 22nd Avenue in a dwelling and garages requiring a second alarm assignment to control it.

OFD Organization Circa 1950

Battalion 2	14th Street & Grove Street
Battalion 3	E. 15th Street & Munson Street
Battalion 4	50th Avenue & Bancroft Avenue
Battalion 5	25th Street & Telegraph Avenue
Battalion 6	51st Street & Telegraph Avenue
Battalion 7	E. 14th Street & 93rd Avenue
Engine 1	14th Street & Grove Street
Engine 2*	1st Street & Broadway
Engine 3	7th Street & Pine Street
Engine 4*	E. 14th Street & 12th Avenue
Engine 5*	Milton Street & Market Street
Engine 6	E. 15th Street & Munson Street
Engine 7	60th Street & Idaho Street
Engine 8	51st Street & Telegraph Avenue
Engine 9*	14th Street & Magnolia Street
Engine 10	Santa Clara Street & Vernon Street
Engine 12*	8th Street & Alice Street
Engine 13	E. 12th Street & 33rd Avenue
Engine 14	Champion Street & Palmetto Street
Engine 15*	25th Street & Telegraph Avenue
Engine 16	13th Avenue & Hopkins Street
Engine 17	Allendale Avenue & Abbey Street
Engine 18	50th Avenue & Bancroft Avenue
Engine 19	College Avenue & Birch Court
Engine 20	E. 14th Street & 93rd Avenue
Engine 21	42nd Street & Montgomery Street
Engine 22	34th Street & Magnolia Street
Engine 23	73rd Avenue & Foothill Boulevard
Engine 24	Moraga Avenue & Thornhill Drive
Engine 25	Butters Drive & Joaquin Miller Road
Engine 26	98th Avenue & Stearns Avenue
Engine 27	96th Avenue & Edes Avenue
Engine 28	Skyline Boulevard & Elverton Drive
Engine 29*	66th Avenue & Fenham Street
Engine 30	1st Street & Broadway
Engine 31	Calaveras Street & Buell Street
Engine 32	14th Street & Grove Street
Engine 33	Skyline Boulevard & Grizzly Peak Road
Truck 1	6th Street & Washington Street
Truck 2	E. 14th Street & 12th Avenue
Truck 3	8th Street & Willow Street
Truck 4	25th Street & Telegraph Avenue
Truck 5	59th Street & San Pablo Avenue
Truck 6	50th Avenue & Bancroft Avenue
Truck 7	E. 14th Street & 93rd Avenue
Truck 8	98th Avenue & Stearns Avenue
Truck 9	Calaveras Street & Buell Street
Fireboat	1st Street & Broadway

* Denotes 2-piece companies (Engine and Hose Wagon)

1960s

The OFD went through a massive restructuring period beginning in the early 1960s. With the help of a local bond measure, 11 new firehouses were constructed between 1960 and 1964. These new stations (5, 6, 12, 13, 16, 17 [31], 19, 21, 23, 27, and 28 [17] replaced outdated and dilapidated quarters that had served the OFD since the horse-drawn or early motorized days. Engines 6, 17, and 21 relocated from the flatlands to better protect the many new neighborhoods being built in the hills of Oakland. Also built in the early part of this decade was a state-of-the-art drill tower, located at Fallon Street and Victory Court. The previous tower had been located at 48th Avenue and MacArthur Blvd., but the construction of the new Hwy 580 (the "MacArthur" Freeway) prompted the OFD to move the training division to its present location.

Station 21

32 Engine at Station 1.

Quad at 24 Engine.

During this decade, older, gasoline-powered engines and trucks were replaced with safer and more efficient diesel apparatus. Also during this period, the Pitman Snorkel, a unique aerial apparatus was purchased by the OFD. Developed by the Chicago Fire Department in the late 1950s, the "Snorkel" was an articulating boom with a platform attached to the tip. It had a pump and supply line and could put two firefighters on the platform in places that traditional aerial ladders could never dream of reaching. The "Snorkel" was a special call apparatus, except on downtown boxes, and was used mostly for its tremendous master stream capability on greater alarm fires.

A landmark change in working conditions for the members of the OFD occurred in the early 1960s with the introduction of the three-platoon system and the 52-hour work week. This was largely due to the diligent efforts of the officers of IAFF Local #55. For the first time in the 90-year history of the OFD, there was now a "C" Shift.

Fires of the 1960s

Tragedy struck the OFD on August 31, 1968, when a fast-moving grass fire claimed the lives of three of Oakland's Bravest. The fire, which began off of MacArthur Blvd. on the Mills College campus, made a run for the road on which Engine 31 was positioned. The private hydrant, which Engine 31 had taken as a water supply, was in terrible condition - the caps and the spindle had been painted over with a heavy coat of paint. The crewmembers at the nozzle, Lieutenant John Carlyon, Hoseman Terry Silveria, and Hoseman Floyd DeMarco were fatally burned when the water tank on Engine 31 ran dry. Mills College was later sued by the widows of the victims and paid almost half a million dollars in restitution to the families of the fallen.

1970s

After the tremendous growth in the 1960s, the OFD faced the first of many budget cuts to come, thanks to Proposition 13. Gone from the OFD was the elite Engine 32, the "Squad" and last of the old Chemical Companies. The first of the Battalion District reductions occurred with the closing of Battalion Chief 6 (at Engine 8) in North Oakland. Fire protection in North Oakland was further reduced with the closing of Engine 7. This company had served the

25th and E. 14th, October 27, 1968.

4th Ave. & East 12th, May 1970.

20th Avenue & E. 21st., July 1971.

Best Fertilizer, May 28, 1972.

as did all engine companies. Due to the budget crisis, except for the three downtown truck companies (Trucks 1, 2, and 4) all of the fire companies now ran with four firefighters, instead of five or six. By late 1979, all the two-piece companies had their hose wagons disbanded.

The removal of the street fire alarm boxes, due to an out of control false alarm rate and excessive maintenance costs, took place in 1978. The box system had served the city for over 100 years. The 9-1-1 Emergency Telephone System began that year and was designed to make reporting emergencies easier, but we soon found out that the system was inefficient when it became overloaded with police business.

Fires of the 1970s

On September 22, 1970, the OFD responded to the largest fire in the East Bay since the 1923 Berkeley Hills Fire burned down to Hearst Avenue. The

425 Foothill Blvd, June 6, 1973.

7th and Campbell, June 14, 1973.

Golden Gate District since 1901, but was considered expendable in the post-Proposition 13 budget crunch. Additional reductions in the command staff occurred with the closing of Battalion Chief 3 in East Oakland. Downtown truck companies used to run with six men. East and North-end truck companies ran with five men,

6211 Telegraph Avenue, March 25, 1973.

Fish Ranch Road Fire started on Marlborough between Norfolk and Grizzly Peak and soon went to six alarms. For the first time in over thirty years, the OFD had to call for mutual aid. Fire departments from as far away as Fairfield and Milpitas responded to Oakland's call for help. Thirty-seven homes were completely destroyed and another 17 homes partially consumed by the flames. Little did those in the OFD know that the Fish Ranch Road Fire was only a precursor to what would be the largest urban-interface fire in U.S. history right here in Oakland just two decades later.

On January 17, 1979, OFD companies were dispatched to the BART Trans-Bay Tube for a "smoke problem" aboard a BART train. First arriving companies found a BART car well involved and many civilians to rescue. After rescuing everyone from the burning train and evacuating them to a waiting train in the opposite bore, these initial companies were overcome by smoke so thick that visibility was non-existent. The gallery, a utility corridor between the two bores, was supposed to be an area of refuge for firefighters. Instead, it became a trap. Lt. William Elliott of Truck 1 was killed and 43 of his brother firefighters were sent to the hospital, some in critical condition.

On the night of July 29, 1979, the OFD fought a fire at the Oakland International Airport that was one of the highest dollar losses in the City of Oakland's history. The loss was conservatively estimated to be in excess of ten million dollars. Six alarms were required to battle the fire in a large hanger, Building 711. Many rare DC-3 aircraft parts were destroyed. The airport's saltwater hydrant system was used for the first time in over ten years to supply the needed gallons per minute for this fire.

OFD Organization Circa 1970

Battalion 2	14th Street & Grove Street
Battalion 3	Derby Street & San Leandro Street
Battalion 4	High Street & Porter Street
Battalion 5	27th Street & Telegraph Avenue
Battalion 6	51st Street & Telegraph Avenue
Battalion 7	93rd Avenue & A Street
Engine 1	14th Street & Grove Street
Engine 2*	1st Street & Broadway
Engine 3	7th Street & Pine Street
Engine 4*	E. 14th Street & 12th Avenue
Engine 5*	34th Street & Market Street
Engine 6	Colton Boulevard & Snake Road
Engine 7	60th Street & Idaho Street
Engine 8	51st Street & Telegraph Avenue
Engine 9*	18th Street & Adeline Street
Engine 10	Santa Clara Street & Vernon Street
Engine 12*	8th Street & Alice Street
Engine 13	Derby Street & San Leandro Street
Engine 14	Champion Street & Palmetto Street
Engine 15*	27th Street & Telegraph Avenue
Engine 16	13th Avenue & Excelsior Avenue
Engine 17	Grass Valley Road & Golf Links Road
Engine 18	50th Avenue & Bancroft Avenue
Engine 19	Miles Avenue & Pressley Way
Engine 20	93rd Avenue & A Street
Engine 21	Skyline Boulevard & Parkridge Drive
Engine 23	73rd Avenue & Foothill Boulevard
Engine 24	Moraga Avenue & Thornhill Drive
Engine 25	Butters Drive & Joaquin Miller Road
Engine 26	98th Avenue & Stearns Avenue
Engine 27	Pardee Drive & Hegenberger Road
Engine 29*	66th Avenue & Fenham Street
Engine 31	High Street & Porter Street
Engine 32	14th Street & Grove Street
Truck 1	14th Street & Grove Street
Truck 2	E. 14th Street & 12th Avenue
Truck 3	7th Street & Pine Street
Truck 4	27th Street & Telegraph Avenue
Truck 5	51st Street & Telegraph Avenue
Truck 6	50th Avenue & Bancroft Avenue
Truck 7	93rd Avenue & A Street
Truck 8	73rd Avenue & Foothill Boulevard
Snorkel 1	14th Street & Grove Street
Fireboat	1st Street & Broadway

* Denotes 2-piece companies (Engine and Hose Wagon)

1980s

The early 1980s saw amazing changes in the racial and gender makeup of the Oakland Fire Department. Firefighting had always been a field traditionally dominated by men. In 1980, the OFD made history as one of the first departments on the West Coast to hire a female firefighter. Several more women soon followed and became members of Oakland's bravest. Another unprecedented event in the history of the OFD occurred on May 1st, 1981. Samuel Golden, an African-American Battalion Chief, was appointed to Chief of Department for the Oakland Fire Department. Chief Golden started his career as a hoseman at Engine 22 and worked his way to the top. Since Chief Golden's tenure, four of the last five Chiefs of Department for the OFD have been African-American.

9 Engine

Budget cuts, begun in the 1970s, continued to plague the OFD and reduce the level of service provided to the citizens. In 1981, five companies were disbanded - Engine 9, Engine 14, Engine 17, Truck 8, and the aging Snorkel. Battalion 5 was also disbanded. In addition to these cuts, the position of Chief's Operator was eliminated.

A new Station 1 was built in 1982 to house headquarters, Engine 1, Truck 1, Chief 2, the Arson Investigation Unit, and a state-of-the-art Fire Dispatch Center. Dispatchers would now use a Computer-Aided Dispatch (CAD) system to transmit alarms to companies, and replaced the antiquated "joker system." With the advent of the CAD, gone now were night watches, reading the tape, pegging the board, and trying to remember all the greater alarm rules that had plagued Oakland firefighters for over 100 years.

The Fire Service began combating an often deadly and unseen enemy during the latter portion of the 19th century. This new threat came in many forms and had to be fought with science and finesse, not water and axes. The Fire Service called this new enemy "Hazardous Materials." In 1981, the Oakland Fire Department established its own Hazardous Materials Team, cross-staffed by specially trained members of Engine 12.

Other changes to the OFD during this decade included installing washing machines and dryers in all firehouses, thus saving the City and taxpayers over $100,000 per year in linen cleaning costs. In addition, truck company numbers were changed to match the station number to which they were assigned. Chief 7 became Chief 3 and Engine 31 was changed back to Engine 17 to aid in the switch to the CAD system.

Fires of the 1980s

Notable fires of the early 1980s included a 5th alarm fire at Abbey Rents, 4244 Broadway. The OFD fought a 5th alarm high-rise fire at 1200 Lakeshore Boulevard on one of the upper-most floors. Yet another 5th alarm fire was fought when a gasoline tanker caught fire on Hwy 880 and High Street. A spectacular 6 alarm fire destroyed the First Methodist Church, located at 24th Street and Broadway in the afternoon of July 1, 1981.

October 17, 1989 was a beautiful fall day, and at

Eng. Randy Kirchner, 1st Haz Mat Tech, 1987.

Hazmat van

From left: Capt. Tony Crudele, Chief Ron Campos and Bryan Honery. The driver of this large dump truck took the corner too fast. He realized he was tipping over and slid out of the driver's seat to the passenger's side. He was not hurt except for a few minor cuts and bruises. The steering wheel was crushed down to six inches of the floor of the cab. He smiled for this picture, he wanted it for his daughter, Nov. 14, 1989. (Paul Leimone Sta. 17)

A semi-tractor trailer came off Freeway 880 and through a house. It took hours to extricate the driver and occupants of the house.

Candlestick Park in San Francisco, the "Battle of the Bay" was set to begin. The pennant-winning Giants were set to face the American League Champion Oakland Athletics in the World Series. At 5:04 p.m., without warning, the entire Bay Area was violently shaken for what seemed like hours, but in fact only lasted fifteen seconds. The Loma Prieta Earthquake, as it was called, registered magnitude 7.1 and caused widespread damage throughout the region. It also killed forty-two people and injured countless thousands. This quake was so powerful that it could be felt in an area of over 400,000 square miles. Oakland was hard hit, with many buildings sustaining structural damage. Several large fires started as a result of broken gas lines. But the most daunting challenge for the Oakland Fire Department was the collapse of the Cypress Freeway. First arriving companies were confronted with a double deck freeway that had collapsed, trapping or killing many drivers under tons of concrete and steel, and starting a number of crushed vehicles to catch fire. Confronted with an almost overwhelming situation, the Oakland Fire Department rose to the occasion and successfully rescued dozens of trapped citizens at great risk to their own lives.

1990s

The 1990s, as in each previous decade, saw the Oakland Fire Department evolve in many ways. Due to several natural and man-made disasters, the Federal Government formed specially trained Urban Search and Rescue Teams throughout the nation. The Oakland Fire Department was chosen to staff one of these highly trained units, California Task Force 4. Now on call for certain periods during the year, the USAR Team can be fully-deployed anywhere in the United States in the event of disaster.

A tragedy occurred in 1990, when Lance Peterson, a firefighter at Engine 4, fell from the apparatus while responding to a fire and was killed. Lance was well-respected as a firefighter, and was known for his great cooking. He was the 11th member of the Oakland Fire Department to die in the line of duty.

In October, 1991, a disastrous fire in the Oakland/Berkeley Hills (called the "Tunnel Fire"), proved to be a catalyst of sorts for the OFD. In its wake, many changes took place. Although the department kept the old VHF radios as a back-up system, a new and improved radio system was installed. This new system continues to enable companies to utilize hundreds of channels, if needed, instead of just four. This primary fireground radio is part of a citywide 800 MHz system the Fire Department shares with a number of other departments (Police, Public Works, etc.). Gone also were the old company numbers that had been an Oakland tradition since its beginnings in 1869. A four-digit number was given to every unit that identified it on a countywide system. Engine 1 became 2541 and Truck 6 became 2576. Over 6,500 fire hydrants in Oakland were retrofitted from 3" discharges to 2-1/2" discharges after mutual aid companies were unable to connect to our hydrants. A Vegetation Management Inspection program was implemented by the Fire Prevention Bureau and carried out by companies. Every spring, all homes and lots in Oakland above Hwy 580, the so-called "threat zone," were targeted for annual inspection. Homeowners are now required to be responsible for making their property fire safe. Yet another change for the OFD as a result of the Hill Fire was the adoption of the Incident Command System (ICS). ICS is now used as a tool by company and chief officers to assist in disaster mitigation.

After the Loma Prieta Earthquake and the Tunnel Fire, a bond measure was passed by taxpayers to ensure that the Oakland Fire Department would be better prepared to face these types of large-scale disasters. Measure I provided funds to purchase a heavy rescue unit, three type III wildland engines, four portable water supply systems (PWSS), or hose tenders, that carry one mile of 5-inch hose each, and a new fireboat, the Seawolf/City of Oakland.

The *Hoga* was retired in 1992, after almost 45 years of faithful service. The Port of Oakland, the fourth busiest in the U.S., warranted a fireboat, and, in 1994, the brand-new Seawolf/City of Oakland was placed in service. Able to pump over 6,500 GPM, she was a tremendous boost to fire protection along Oakland's waterfront.

Seismic retrofitting of all Oakland firehouses began in the mid-1990s to ensure the structural stability of our buildings. Also installed in every firehouse during the retrofit was an emergency generator, which could power critical systems such as computers and apparatus doors, in the event of another major earthquake. All Oakland firehouses were retrofitted with separate female bathrooms to accommodate the many women that had recently been hired.

Several new firehouses were also built during the

Greg Williams at 82nd Ave. and San Leandro St., Big Rig Truck Stop, 1997. (Courtesy of Paul G. Fellows)

82nd Ave. and San Leandro St., Big Rig Truck Stop, 1997. (Courtesy of Paul G. Fellows)

Jennifer Schmid (left) and Annette Goodfriend (right) at Station 17 after a 4-alarm fire. (Courtesy of Annette Goodfriend)

mid-1990s. The Hazardous Materials Team moved from Engine 12 to Engine 3 and Truck 3's new firehouse on 14th and Peralta Streets in West Oakland. A new station was built for Engine 20 and Truck 7 on 98th Avenue and International Boulevard (old E. 14th Street). Engine 26 moved into new quarters across the street from their old quarters at the top of 98th Avenue. After being quartered with Engine 25 for almost four years (due to the Gingerbread Firehouse being damaged in the Loma Prieta Earthquake), Engine 24 finally moved into their new quarters in 1997. The Oakland Fire Department took over fire protection for the Oakland International Airport from the Port of Oakland in 1998. A massive new fire station (numbered Engine 22) was built for the huge Aircraft Rescue and Firefighting (ARFF) units housed there.

Several companies were re-opened during this time. Engine 7, which had previously served in North Oakland, was now in Oakland's North Hills providing that area with much-needed fire protection. The new firehouse is located near where the "Tunnel Fire" started. Re-opened in the Grass Valley area was newly renumbered Engine 28.

During this time, the OFD began an aggressive apparatus acquisition in order to replace a very aging fleet of engines and trucks. As late as 1998, some engine companies were responding with firefighters riding on the tailboard because the older rigs did not have jump seats. The apparatus committee was able to purchase new state-of-the-art Pierce Quantum engines and trucks, a huge improvement over what was currently in use.

During the latter part of the decade, all Oakland firefighters, who had previously been trained as First Responders, were now required to maintain Emergency Medical Technician-D certification. Also during this time, the OFD started to train some of its own personnel as paramedics. This advanced life support capability has had a very positive impact on our ability to save the lives of the people we serve.

Fires of the 1990s

On the morning of October 20, 1991, several Oakland engine companies were mopping up a grass fire near the Caldecott Tunnel that had occurred on the previous day. The fire had been a small one. No structures had been damaged and no one had been injured. While mopping up, a stray ember that had lingered all night was blown into some dry brush by a surprisingly strong gust of hot wind. The brush ignited and instantaneously it seemed as if the entire hillside was ablaze. Within minutes, flames were racing into dry stands of eucalyptus. Embers were blown across eight lanes of Hwy 24 starting spot fires everywhere. Before the day was through, the "Tunnel Fire" would tax the members of the OFD to their

Paul Fellows on 6 Truck, California Spa Fire, August 1999. (Courtesy Paul G. Fellows)

California Spa Fire, August 1999, 18 E-6 T, 4-alarms, High St. Exit/Coliseum. (Courtesy of Paul G. Fellows)

Rick Young, John Larson, Mark Fraser mop up after the Oakland Hills fire, 1991. (Couresty of S. Miguel)

Broadway Fire Jan. 10, 1999. Cedric Price, Aaron Montes, Olivia Moore (sitting), Joseph Torres, Leon Primas (sitting), Jeanne Andrews moments after the loss of Tracy Toomey. (San Francisco Chronicle, photographer Michael Macor. Courtesy of Joseph Torres.)

fire was reported at mid-morning, downtown companies rushed to the scene. The fire went to several alarms and looked to be under control when disaster stuck. Tracy Toomey, a veteran firefighter working on Engine 12 that day, was killed when the second floor collapsed on him. Also injured in the collapse

The Tracy Toomey fire.

We will never forget Tracy.

IN MEMORY OF
FIREMAN TRACY TOOMEY
WHO ON JANUARY 10TH, 1999 MADE THE ULTIMATE SACRIFICE FOR THE CITIZENS OF OAKLAND. A LOVING HUSBAND AND FATHER, A DEDICATED 27-YEAR OAKLAND FIREFIGHTER, A MARINE CORPS VETERAN, A BLACKSMITH ARTIST, AND, MOST OF ALL, OUR FRIEND. HE WILL BE MISSED BY ALL.

Our friend and brother.

very limits. It would be three days before the fire was declared under control. Twenty-five people would lose their lives, including Battalion 4, Chief James Riley. One hundred fifty people would be injured, some of them critically. The fire burned almost 1600 acres and destroyed 2700 structures. The fire burned so hot and so fast that it is estimated that a structure burned every 11 seconds. This was truly an "urban firestorm" of the greatest magnitude. Mutual aid came from all over California – over 400 engines responded. The cost of this fire was beyond all measure. However, the dollar loss of the Tunnel Fire came to $1.68 billion.

A fire occurred on the morning of January 10, 1999, at an old bar called the Back on Broadway. The building had been significantly modified over the years and was no longer structurally sound. When a

Old Federal Building, 4-alarms, fire on roof, 2000. (Courtesy of Paul G. Fellows.)

MacArthur across from Mills College housefire, 18 E - 6 T, 2001. (Courtesy of Paul G. Fellows.)

Car fire, Station 23, FF Keith Hall.

The Twenty-first Century

A new chapter for the OFD began on May 30, 2000, when Engine 23 became the first ALS (Advanced Life Support) engine company in Oakland. The era of Para-medicine was upon us. By the year 2002, all but one OFD engine company would carry a Paramedic.

A new Emergency Operations Center (EOC) was built at Station 1 in 1999. The EOC was designed to be a command center for the city in the event of a major crisis. It can be staffed with fire department and police officials, public works employees, dispatchers, public information officers, and officials from outside agencies, such as the U.S. Coast Guard, the FBI, and ATF.

After the terrorist attack on the Twin Towers in New York City on September 11, 2001, Oakland's USAR Team (CATF – 4) was deployed to aid in the recovery efforts. Training in the field of "weapons of mass destruction" had become a very important part of being a firefighter. It could happen in Oakland.

The crews of Engine 8 and Truck 5 moved into a beautiful new firehouse on the site of their old quarters in December, 2002. This new firehouse replaced a dilapidated station built in 1950 as temporary quarters.

Two new American-LaFrance aerial ladder trucks were purchased and placed into service in 2003 at Truck 1 and Truck 4. The older Pierce aerials used by these companies went to Truck 3 and Truck 5. All seven truck companies received new Hurst tools and power units, in some cases replacing first-generation tools from the 1970s.

Going Forward

In 2003, budget cuts at the state and city level had a tremendous impact on the OFD. Gone were the Fireboat Seawolf and Engine 2. In addition to this loss, two engine companies at double houses were closed daily on a rotational basis. The loss of these three engine companies undoubtedly had an impact on our ability to combat fires and to save the lives and property of the citizens who count on us.

The OFD has a long and proud history. Some who served gave all, and all who served gave some. For the OFD, there are countless stories written, told or forgotten, and many more stories waiting to be told.

were Lieutenant Kenny VanGorder and Firefighter Sheree Rodriguez, also of Engine 12. Due to the heroic efforts of Truck 1, all three of the trapped firefighters were pulled from the rubble, but not before Firefighter Toomey perished.

OFD Organization - Present

Battalion 2	16th Street & Martin Luther King, Jr. Way
Battalion 3	98th Avenue & International Boulevard
Battalion 4	High Street & Porter Street
Engine 1*	16th Street & Martin Luther King, Jr. Way
Engine 3	14th Street & Peralta Street
Engine 4*	International Boulevard & 12th Avenue
Engine 5	34th Street & Market Street
Engine 6	Colton Boulevard & Snake Road
Engine 7	Alvarado Road & Amito Avenue
Engine 8*	51st Street & Telegraph Avenue
Engine 10	Santa Clara Street & Vernon Street
Engine 12	8th Street & Alice Street
Engine 13	Derby Street & San Leandro Street
Engine 15*	27th Street & Telegraph Avenue
Engine 16	13th Avenue & Excelsior Avenue
Engine 17	High Street & Porter Street
Engine 18*	50th Avenue & Bancroft Avenue
Engine 19	Miles Avenue & Pressley Way
Engine 20*	98th Avenue & International Boulevard
Engine 21	Skyline Boulevard & Parkridge Drive
Engine 22	751 Air Cargo Road 5 ARFF Units
Engine 23	73rd Avenue & Foothill Boulevard
Engine 24	Snake Road & Moraga Avenue
Engine 25	Butters Drive & Joaquin Miller Road
Engine 26	98th Avenue & Cherokee Avenue
Engine 27	Pardee Drive & Hegenberger Road
Engine 28	Grass Valley Road & Golf Links Road
Engine 29	66th Avenue & Fenham Street
Truck 1	16th Street & Martin Luther King, Jr. Way
Truck 2	E. 14th Street & 12th Avenue
Truck 3	7th Street & Pine Street
Truck 4	27th Street & Telegraph Avenue
Truck 5	51st Street & Telegraph Avenue
Truck 6	50th Avenue & Bancroft Avenue
Truck 7	98th Avenue & International Boulevard
Rescue 1	16th Street & Martin Luther King, Jr. Way staffed by Truck 1
HazMat 1	14th Street & Peralta Street staffed by Engine 3

* Currently two of these engine companies are closed daily on a rotational basis as part of budget cuts.

Contributors to this historical synopsis of the Oakland Fire Department include Asst. Chief Neil Honeycutt (OFD, retired), Lt. Ed Clausen (OFD, retired), Clarence Crum (OFD, deceased), and Capt. Geoffrey Hunter (OFD, active). Thanks to all.

Sept. 11, 2001, from left: FF Emmett Fahey (4Tk), Lt. Marshall McKee (15 Eng.), FF Roger Prieto (8 Eng.). Crews returned to quarters after a structure fire, to see the vicious attacks of the World Trade Center. (Courtesy of Roger Prieto)

May 2, 2001, Foothill and Seminary Ave., 4-alarms, 18 - 6 T. (Courtesy of Paul G. Fellows)

Haz Mat, FF Dan Keenan, Eng. Perry Washington, Dec. 2001. (Courtesy of Keith Hall)

Fire Chiefs of the Oakland Fire Department

Oakland Volunteer Fire Department
1853 - 1869   John W. Scott

Oakland Fire Department

1869 - 1870	John C. Nalley
1870 - 1872	Miles Doody
1872 - 1874	George Taylor
1874 - 1877	Matthew de la Montanya
1877 - 1878	Fred O. Fuller
1878 - 1883	James Hill
1883 - 1889	James Moffitt
1889 - 1893	James F. Kennedy
1893 - 1896	Elburton B. Lawton
1896 - 1898	William H. Fair
1898 - 1915	Nicholas A. Ball
1915 - 1921	Elliott Whitehead
1921 - 1927	Samuel H. Short
1927 - 1947	William G. Lutkey
1947 - 1955	James H. Burke
1955 - 1972	James J. Sweeney, Jr.
1972 - 1976	Stephen L. Menietti
1976 - 1980	William L. Moore
1980 - 1981	Walter J. Pierson (Interim Fire Chief)
1981 - 1986	Samuel L. Golden
1986 - 1991	Godwin G. Taylor
1991 - 1996	P. Lamont Ewell
1996 - 1999	John K. Baker
1999 - 2004	Gerald A. Simon
2004 - Present	Daniel D. Farrell

40

Aircraft Rescue Fire Fighting

Replacement OFD "ARFF" Airport Fire Station #22 building M-911 opened for duty September 1998, 11 million dollars, 30,000 sq. ft., dedicated to the historic first fully African American Staffed OFD Fire Station #22.

OFD Engines circa early 1970 for fatal aircraft fuel tanker fire north airport hangar 4.

Rescue 5, 22 tons, 60 gallons diesel, 1,500 gal. water, 210 gal. AFFF, 450 lbs. dry chemical PKP, cost, $239,000, 0-65 mph 24 seconds.
UPS Boeing 747-300, 450 tons, 58,000 gallons Jet-A fuel, $312,000,000, 0-578mph 480+seconds,

"The One That Got Away" XFL Football promotional blimp broke loose from mooring at Oakland Airport North Field and landed 5 miles away at 5th and Embarcadaro Cove Marina.

October 1999 ARFF Station 22 personnel with Station 27 crew for "Fleet Week" F-18 Hornet A/C familiarization. From left: James Harris, Neil Gadison, Glenn Berry, Kruger Story, Julie Green, Paul Idle, Mike Hill, Blue Angel Crewman, Ron Bragg, Pete Martinez, Blue Angel Crewman

October 1999. A combined gathering of all ARFF apparatus at the United Maintenance Hangar for fleet week, Blue Angel F-18 Hornet aircraft familiarization. Rescue (s) 2-8 15,020 gal. water, 1,590 gal. AFFF, 1,500 lbs. dry chemical PKP.

Antique meets future. Vintage WW II 1945 Boeing B-29 "Superfortress" along side 1993 Oshkosh T-1500 at Alaska Maintenance Hangar.

41

OPD Argus MD-500 D police helicopter taking off from ARFF Station 22.

OFD ARFF Mike Hill doing a daily "Apparatus Check" of Rescue 6.

Last of the "Chubbs." Larry Rosso's favorite apparatus. 1972 Chubb Pathfinder. New: $270,000, 3,500 gal. of water, 500 gal. of AFFF, 41 tons, 1,800 GPM, 70 mph.

FF G. Stein OFD with Cub Scout Troop 502 next to Rescue 7 "Snozzle." $750,000, 3,000 gal. water, 420 gal. AFFF, 900 GPM.

Rescue 2 "Oshkosh" T-3---, 2002 Model, $460,000, 3,000 gal. water, 420 gal. AFFF, 450 lbs. dry chemical PKP. "A" Shift Crew, from left: FF Darrin White, FF Stan Killingsworth, CA Jim Smith, FF Paul Budiao, FF Senator Brunson, FF Ernie Sherman.

Salt Lake City, Utah ARFF Training Facility for mandated ARFF "Live Burn" annual training.

"Extend the Ladder Drill."

42

FIRE SAFETY AND EDUCATION DAY (WEEK)

The Oakland Fire Department Fire Prevention Bureau, in partnership with Blue Cross of California and State Farm Insurance, initiated an annual Fire Safety and Education Day on October 14, 1997. This annual event for children in kindergarten through fifth grade was designed to help eliminate fire deaths among Oakland children by teaching them how to prevent fires in the home and how to stay safe in a fire emergency.

In response to a series of children's fire deaths in the city of Oakland, the Fire Prevention Bureau recognized a need for better public education, which led to the creation of the fire safety and education day. A task force was organized to develop specific guidelines and processes to create an annual education experience for the children. Fire Department efforts, spearheaded by former Fire Marshal Jerry Blueford and Inspector Gilbert Cody, have presented an opportunity to make fire prevention inroads and fire safety efforts by children.

Program benefits include the following:

1) Established a partnership with the Oakland Unified School District to add Fire Safety and Education Day to the school's curriculum;

2) Established a partnership with Blue Cross of California to provide 2,000 free smoke detectors for distribution;

3) Established a partnership with State Farm In-

After the end of the day hundreds of kids have passed through. From left: Kim Catano, Gil Cody, Joan Austin-Garrett, Camille Rodgers.

From left: Guy Fraker, State Farm Insurance; Councilmember Larry Reid; Odessa Bolton, Blue Cross of California; former Fire Marshal Jerry Blueford.

Stephanie Massey with Domino (dog in background) showing the children how to "Stop, Drop and Roll."

From right: Former Fire Marshal Jerry Blueford, Fire Chief Gerald Simon, Fire Fighters John Reinthaler, Rick Chew, Mickey Quinn.

Chris Clark and Jim Howell preparing lunch.

surance to provide fire safety information brochures, batteries, smoke detectors and t-shirts;

4) Provides an opportunity for public education outreach by using the Fire Safety House to point out potential hazards in the home and how they can avoid possible injury to themselves or family members. Some of the safety features include releasable bars on windows, fireplace hazards, how to crawl through simulated smoke-filled corridors and other hands-on safety practices so that children may see and participate in a controlled and safe learning environment;

5) Provides public education outreach by using the "Stop, Drop and Roll" program that allows children to practice what they should do in the event of a fire situation;

6) Provides an opportunity for City Officials and staff to directly interact with children to reinforce the importance of fire safety and fire prevention measures.

All materials, prizes and food were donated by private businesses.

Children crawling through the Fire Safety House.

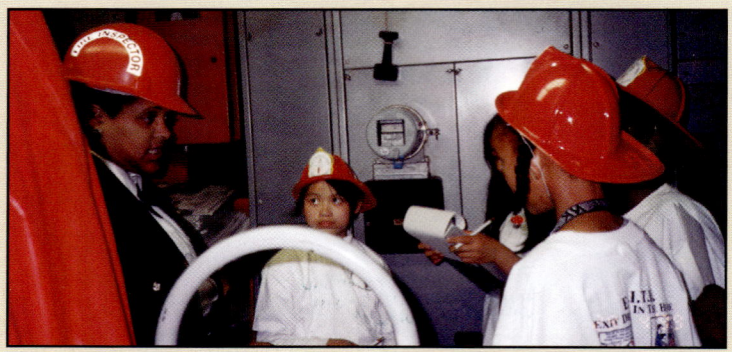

Sylvia Chaney-Williamson and the Junior Fire Marshals

Fire Fighters assisting kids in line to inspect the Fire Safety House.

The children gathered for presentations and awards.

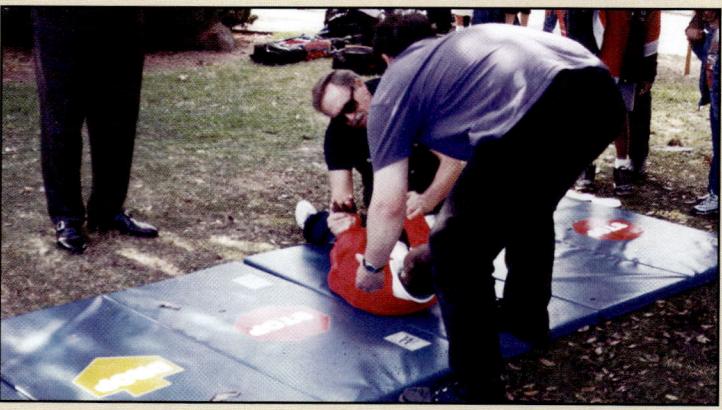

Fire Fighters assisting kids with how to "Stop, Drop and Roll."

Security bar demonstration by Jim Howell.

Another "at the end of the day" group: Gary Collins, Mike McCarthy, Chris Clark, and Kim Catano.

Oakland Firefighters
Creating a positive difference in the lives of individuals through Random Acts of Kindness

Oakland Firefighters Random Acts Initiative

Every year firefighters across the nation respond to millions of 911 calls. Many involve a death, the destruction of personal property, or other damage to the fabric of our communities. We witness people who are suddenly faced with tests to the very limits of their bravery and endurance, sometimes heroically and sometimes tragically. Over the years we firefighters have found ourselves subjected to some physical and emotional stress in the course of our service, but we continue to respond in a professional way and support our side of society's safety net.

The Oakland Firefighters Random Acts Initiative has been founded by individual firefighters who truly believe we can make a positive difference in other ways besides our emergency response work. We are a registered 501(c)(3) tax exempt charity with a simple aim – to help our fellow firefighters give something extra back to the communities they live in and care for.

With a board made up of diverse individuals of many ranks, serving on a purely voluntary basis, we aim to fulfill our mission statement: "Creating a positive difference in the lives of individuals through Random Acts of Kindness". Our motto "No Egos, Badges or Resumes" means simply that the program is for everyone in our department, and all we want is a commitment from the heart – not a reflection of ego, status or ambition. Because everyday firefighting brings us up close with the lives of people who truly need a hand to get by, we provide a channel for our colleagues to bring donations or their requests for help. A central location has been set up to handle suggestions and donations, and our board has heard requests for such items as a wheelchair for a senior citizen, a refrigerator for a family and a computer for a disabled child.

Our grassroots organization is also an outlet for firefighters to get support for their own outreach initiatives and interests. We have a former Oakland school teacher who has created a Random Acts of Reading group for elementary schools. Working through the Random Acts network, he has been gathering volunteers and reading to children from Oakland's inner city areas. Others have brought their passion for helping children with disabilities, or burn victims, and found Random Acts a source of practical and financial support.

This year we also created a Citizen Award, a way to identify and celebrate silent citizen heroes who have gone to extraordinary lengths to help others in trouble or at risk. The thinking behind a program like this is that, as we return from our 60,000-plus 911 calls around Oakland each year, firefighters can be empowered to plant positive seeds in their districts and create lasting memories for themselves. Human interaction that produces laughter or tears of joy is a fine medicine for us and our community.

We invite you to join our supporters, including the staff of the San Francisco office of news and information company Reuters America and many individuals. With a donation your organization can support a program that is a clear win for all involved. Contact us for more information or a presentation by phone at 510-597-5034, or write us at: Oakland Firefighters Random Acts, PO Box 18842, Oakland, CA 94619. Our Federal Tax ID number is 94 338 7723. Please make checks out to: Oakland Firefighters Random Acts. Know that all funds will go directly to creating Random Acts of Kindness.

Larry Hendricks
Director

P.O. Box 18842 - Oakland, CA 94619 - www.ofrandomacts.org - 510-597-5034

Office of Emergency Services (OES)

The Oakland Fire Department's Office of Emergency Services (OES) is responsible for the city's emergency management organization including disaster planning, response, recovery, and mitigation programs.

Among the emergency management programs that OES currently supports are:
- Citizens of Oakland Respond to Emergencies (CORE), neighborhood preparedness program that corresponds to the national Community Emergency Response Team (CERT) program
- Oakland Radio Communication Association (ORCA), a ham radio club
- Alerting & warning sirens
- Emergency Operations Center (EOC), which is activated in response to disasters and is located at 1605 Martin Luther King Jr. Way
- Emergency Management Board, which serves as Oakland's Disaster Council

The Office of Emergency Services also manages hazardous materials programs. Oakland is one of sixteen California cities that is Certified Unified Program Agency (CUPA). The mission of the Unified Program is to protect public health and safety, to restore and enhance environmental quality, and to sustain economic vitality through effective and efficient implementation of the Unified Program.

The Unified Program (UP) was created by California Senate Bill 1082 (1993) to consolidate, coordinate, and make consistent the administrative requirements, permits, inspections, and enforcement activities for the following environmental and emergency management programs:

- Hazardous Materials Release Response Plans and Inventories (Business Plans)
- California Accidental Release Prevention (CalARP) Program
- Underground Storage Tank Program
- Aboveground Petroleum Storage Act Requirements for Spill Prevention, Control and Countermeasure (SPCC) Plans
- Hazardous Waste Generator and Onsite Hazardous Waste Treatment (tiered permitting) Programs
- California Uniform Fire Code: Hazardous Material Management Plans and Hazardous Material Inventory Statements

Oakland Black Firefighters Association

AFFILIATIONS NAACP LIFE MEMBER, INTERNATIONAL ASSOCIATION BLACK PROFESSIONAL FIREFIGHTERS

History of the Oakland Black Firefighters Association

The Oakland Black Firefighters Association was organized in December 1973 and today has a membership of more than 140.

Many of the founders of the OBFFA entered the Department when the assignment policy of new members was based strictly on one's ethnic origin. Specifically, if the member was identifiably Black, he was assigned to 22 Engine, then located on Magnolia Street near 34th Street. This policy originated the day the Black firefighters were hired on January 1, 1920. These original Black members of the Department were aware of the disadvantages and degrading aspects of this policy. However, the tenor of the times dictated that they accept it or resign.

Through the efforts of the National Association for the Advancement of Colored People (NAACP) and Black community leaders the Oakland Fire Department was integrated in May 1953. The Black members of the department, then numbering around thirty, were placed in several stations throughout the city. While these changes within the department were generally enacted without major incident, some Blacks encountered strong opposition and harassment in their new assignments.

Over the ensuing years the Black members of the Department maintained contact with each other to compare experiences, recruit for entrance examinations and to attend occasional retirement functions or funerals for the original members of 22 Engine. These contacts, and the sporadic entry of more Blacks into the department, eventually pointed to a need for a formal organization.

The basic goals of the Oakland Black Firefighters Association were recruitment, training, and the encouragement of friends and relatives to make the fire department a career. Toward these goals, several Black members conducted classes in private homes, various churches and other public sites during the mid sixties and early seventies.

The Oakland Fire Department had a total of forty-seven (47) Black members when the Association was organized in 1973. Today, Black members represent approximately one fourth of the Oakland Fire Department, some 140 strong. In the area of promotions among Blacks in the department, the statistics also reveal some dramatic and positive changes. The OBFFA feels that our presence has been an effective factor in many of these advancements and we take great pride in contributing to this progress.

Another primary function of the organization has been involvement in a variety of community outreach projects. These include college scholarships, holiday food baskets for the needy, and a Youth Fire Academy. Additionally, OBFFA has demonstrated strong community support through the financial backing of numerous youth and senior health and educational programs throughout Oakland. OBFFA is also a Life Member of the NAACP and many other community-based organizations.

In conclusion, membership is open to any member of the Oakland Fire Department who genuinely supports the goals and objectives of the Oakland Black Firefighter's Association.

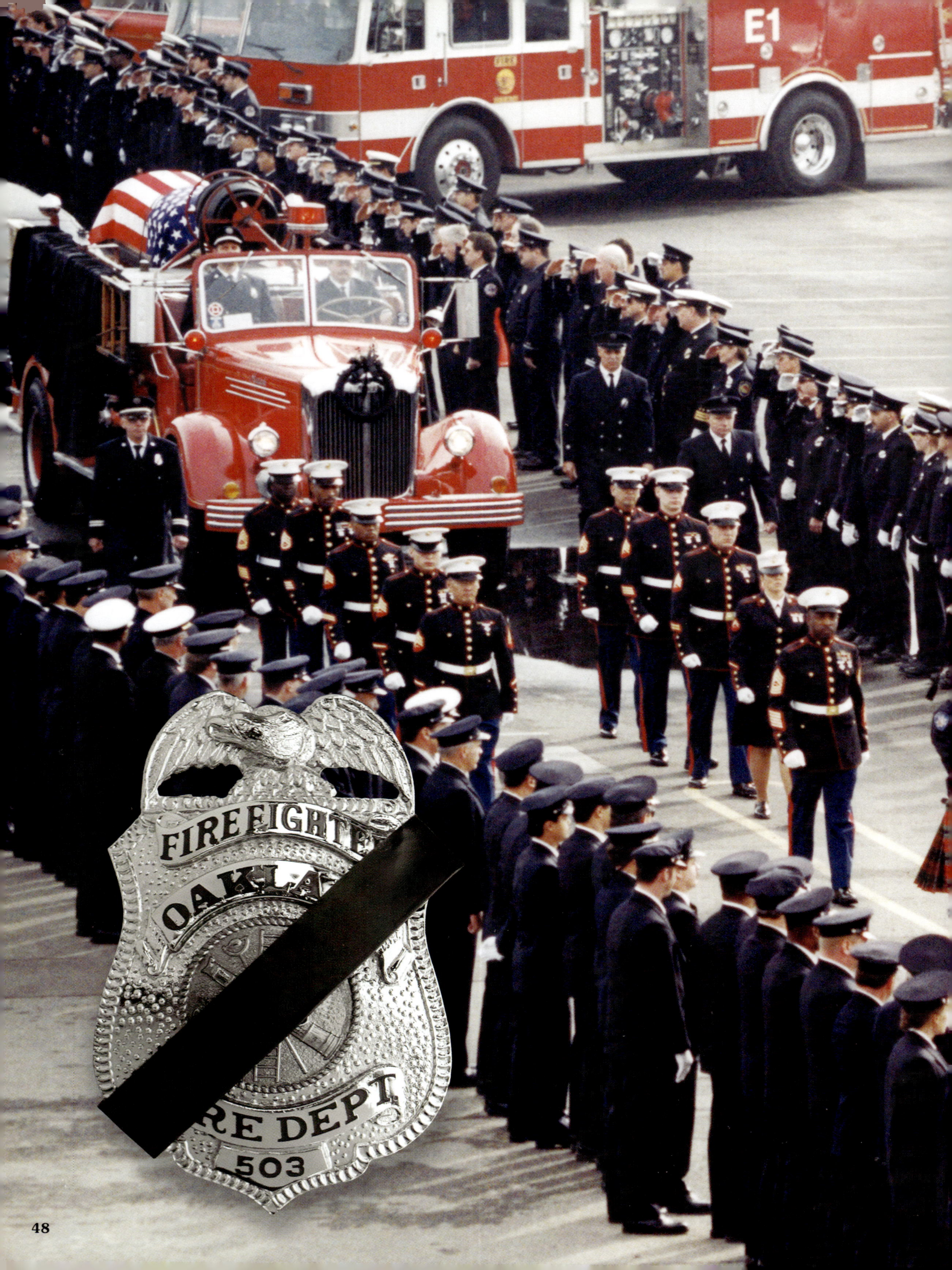
E1
FIRE
FIREFIGHTE
OAKLA
RE DEPT
503

LINE OF DUTY DEATHS FOR THE OAKLAND FIRE DEPARTMENT
OFD Firefighters who gave the supreme sacrifice in the performance of their duties.

Year	Name	Unit	Cause
1903	Hoseman Frank Parker	Engine 3	Building Collapse
1925	Hoseman Thomas Heelan	Engine 1	Train Collision
1940	Hoseman Dewey Records	Chemical 1	Building Fire
1946	Captain Joseph Pimentel	Engine 24	Engine Accident
1947	Hoseman Louis Cetraro	Engine 3	Roof Collapse
1957	Lieutenant Ralph Waalkes	Engine 32	Electrocution
1968	Lieutenant John Carlyon	Engine 31	Wildland Fire
1968	Hoseman Floyd DeMarco	Engine 31	Wildland Fire
1968	Hoseman Terry Silveria	Engine 31	Wildland Fire
1979	Lieutenant William Elliott	Truck 1	BART Fire
1990	Firefighter Lance Peterson	Engine 4	Fell From Apparatus
1991	Battalion Chief James Riley	Battalion 4	Wildland Fire
1999	Firefighter Tracy Toomey	Engine 12	Building Collapse

Oakland Fire Department Personnel

OFD Recruit Class 8-90

Doug J. Abbott	 Susan L. Abel	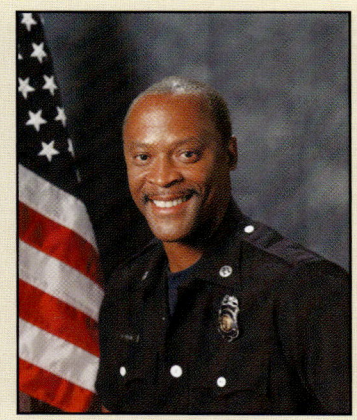 Douglas S. Abram	 Chuck Accurso
 Jake Adams	 Lenroy Adams	 Edward M. Aff	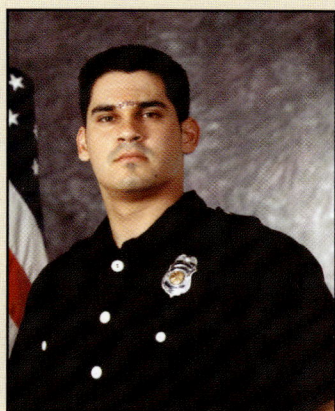 Richard Aguilar
 Matt B. Amormino	 Paul Anderson	 Jeanne L. Andrews	 Melvin Andrews
 Nerissa Andrion	 Dominic R. Antes	 Alex E. Archuleta	 Cruz Arellanes Jr.

Darryl J. Ashley	 Lisa J. Askew	 John M. Asport	 Joan Austin-Garrett
 Charles A. Baker	 Lisa A. Baker	 Damian Bala	 Joanelle Bala
 Raymond V. Banks	 Sherri Banks	 Phillip Basada	 Robert D. Beckley
 Lamont B. Becton	 Benjamin L. Beede	 Gabriela B. Beldiman	 Coleen Bell

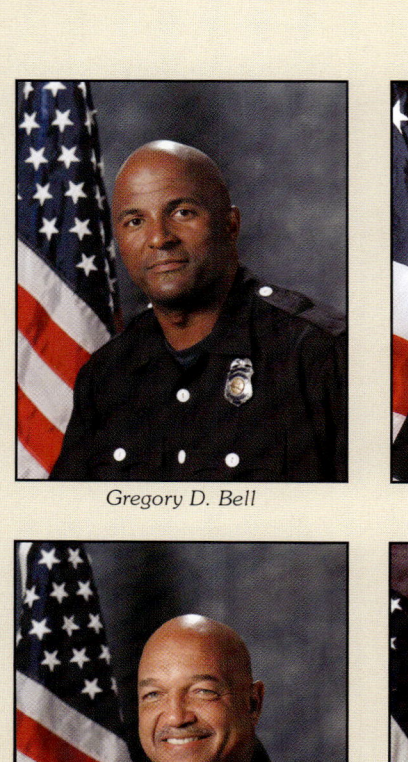

Gregory D. Bell	Phillip Bell	Tanisha B. Bernard	Erick D. Berry
Glenn D. Berry	Roger E. Bird	Jerry E. Blueford II	Eleanor M. Bolin-Chew
Neptali B. Bonifacio	Kiethwin L. Bowers	James Bowron	Donald R. Bozman Jr.
Marlon Brandle	John Ryan Brierly	Daphne Briscoe	Sean Brown

Nathaniel Brownlow	David A. Brue	Bryan Brumfield	Senator Brunson
Paul B. Budiao	Linda Buell	Barbara Burgueno	Betty Burns
James Calhoun	Kathleen Campbell	Ronald K. Carter	Steven D. Carter
Gregory J. Case	Mario Castillo	James R. Catalano	Kim Catano

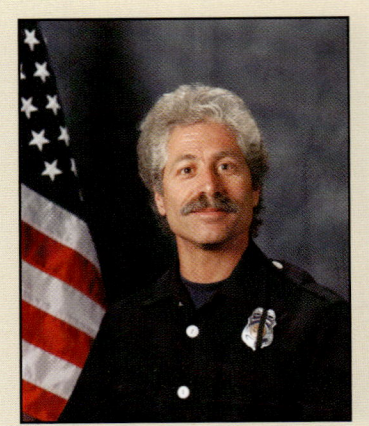

Joe Caulfield

Lucien Cazenave

Denny Chan

Sylvia Chaney-Williamson

André B. Chapital

David E. Chew

Richard Chew

Steven Chew

Gilberto Chiguila

Gene Chin

Tracey Ann Rene Chin

Chris T. Clark

Weldon Clemons

Gilbert M. Cody

John A. Coffer

Christopher Colgan

Gary C. Collins	 Julian Comeaux	 Jay Comella	 Jesse R. Comstock
 Ritalinda Concepcion	 Gary K. Conley	 James Conley	 John Connelly
 Kenneth M. Costa	 Damon Covington	 Bonnie Cox	 Pablo Cruz
 Victor M. Cuevas	 Carolyn Culberson	 Kenneth Cuneo	 Jacqui Curtis

Chris B. Dalrymple	 Steve Danziger	 Heather Dawkins	 Rich De Glymes
 James R. De Lacy	 Elaine De Silva	 Kenneth Delgado	 Hanns Detlefsen
 William D. Detlefsen	 Anthony C. Di Stefano	 Willie B. Dixon	 Victor Dizon
 John M. Dolan	 Renee Domingo	 Tracy Donaldson	 Michael D. Donner

57

Mark C. Douglas	 William B. Douglas	 Eric Draper	 Melinda Ann Drayton
 Glenn Dunham	 James W. Duvall	 Daniel T. Dwyer	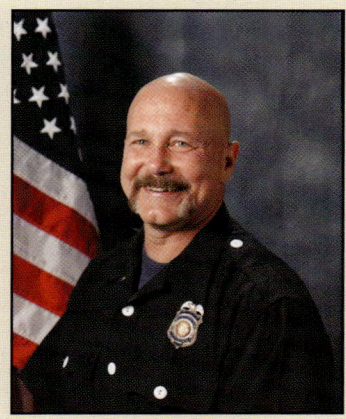 Timothy H. Eade
 Russell J. Earle	 James D. Edwards	 David Eger	 Thomas Elento
 Justin Elliott	 Lorenzo Ellison	 Jackie Emerson	 Kurt Emke

Harold Ray Epps	Kenny Ergun	Kenneth B. Erhardt	David Espino
Keith A. Evans	Emmett J. Fahey	Helen Fairley-Summers	Daniel D. Farrell
John Farrell	Richard Feigel-Perez	Paul Fellows	John M. Ferguson
Mark D. Ferreira	Moses Ferrer	Sergei Fesai	Keenan Fincher

Keith E. Flashberger	 Christopher A. Foley	 Bruce Fontelera	 Debbie Ford
 Annette Fountaine	 Zachary W. Fraser	 Mark A. Fraticelli	 Lorenzo S. Frediani
 George T. Freelen	 Charlotte Fujii	 Harold N. Gadison	 Thomas D. Gallinatti
 Charles A. Garcia	 Dwight Garcia	 Philip Garcia	 Robert Garcia

Monty J. Gardea	Carl Gardner	Sean Edward Gascie	Darryl N. Gaskin
Alan Gaul	Neil Gentry	Nicko Georgatos	Charles Glass
Hernán Gomez	Scott Gonsolin	Bobby Gonzalez	Glenn S. Gonzalez
Annette Goodfriend	Dwayne Gray	Julie E. Green	Kim M. Green

61

Jason A. Greengrass  | George G. Gregoire 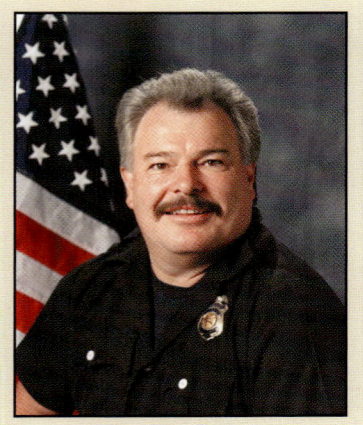 | Roland R. Gregoire  | Chuck Gresher

Leroy Griffin  | Ronald E. Griggs Jr.  | Sylvester Grisby  | James A. Gross

Carlos A. Grunwaldt  | Harvell Guiton  | Gordon F. Gullett  | Derek Hajny

Keith Michael Hall  | Michael D. Hall  | David K. Halliday  | James Halpin

62

Connie Hammond	Bradley L. Harger	Donnie J. Harris	Gregory J. Harris
Yvette Harris	Carlos Harvey	Jerome Hathaway	James Patrick Haughy II
Jeffrey Haughy	Chris R. Heath	David E. Hector	Christen Heinrich
Scott G. Hellige	Larry Hendricks	Michael Hill	Michael Hillesheim

Zachary C. Hilton	David Hines	Howard Hoard	Mark H. Hoffmann
Derek A. Hogerheide	Joseph Hoke	Jacob A. Holmes	Valida D. Holmes
Howard C. Holt	Donna Hom	Lawrence S. Hom	Michael A. Hoskins
Mikhail Hoskins	Richardine Howard	Geoff Hunter	Eduardo H. Ibarra

Paul T. Idle	 John Irwin	 Anthony R. Jackson	 Johney L. Jackson Jr.
 Wellington Jackson	 Vibhor Jain	 Larry James	 Richard James
 William Jarrett	 Porya Jeddi	 Ann Marie Jensen	 Shad D. Jessel
 Ronald Johnson	 Keith L. Jones	 Sherman Jones	 James A. Joseph

65

Leonard T. Jung  | Coy M. Justice    | Malik H. Kafele  | Margaret Kaigler-Armstead

Lee D. Katsanos  | Daniel J. Keenan  | Mary Keeton  | Kevin M. Kennedy

Rusty Keyes  | Fitzroy E. Killingsworth  | Stanley E. Killingsworth  | Edward J. Kilmartin

Betsy Kimmel  | Donna King  | Randy E. Kirchner  | Denise Kittel-Nwuke

Gary C. Klingler	Mathew Knaus	Jamie M. Knudsen	Arosh Kouhi
Rebecca Kozak	John D. Kuehl	Ralph W. Kurio	Sean D. Laffen
Father Jayson Landeza	Christopher Landry	Dwight Langford	Richard C. Lapora
John Larson	Shonda Leary	Dennis Legear	Paul J. Leimone

67

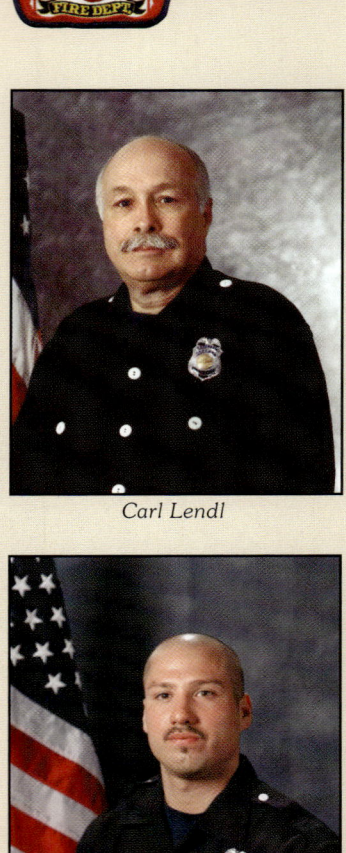

Carl Lendl	 Frank R. Leslie	 Daryl Liggins	 Robert Lipp
Alvaro Lizama	 Edward Llamas	 Jason A. Lloyd	Erik N. Logan
 Lisa Logan	 Brian J. Lopes	 Nickolas Luby	 David MacPherson
 Mark Maddox	 Perry B. Magill	 Lester Mahoney Jr.	 Santalynda Marrero

Dan Marshall	Nola Marshall	Thomas Marshall	Janet Martin
Jeromie Martinez	Peter G. Martinez	Ronald Martinez	Walter Martinez
Charles Mathis	Keith Matthews	James A. Mc Carty	Scott McCrary
Marshall S. McKee	James McNicholas	Ian McWhorter	Randy L. Medina

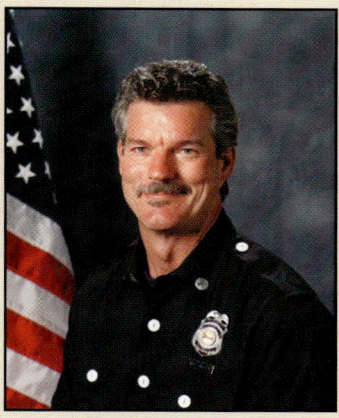

Mark Meiers	 Ryan A. Meineke	 Manuel Mejia	 Dennis R. Menge
 Christian Mercado	 Ahvrumm Merritt	 Stephen Miguel	 Roy A. Miles
 Earl Lee Miller	 June Miller	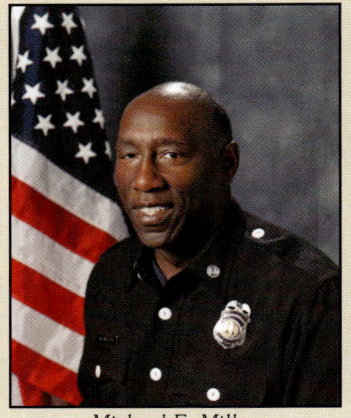 Michael E. Miller	 Maurice R. Miranda
 Thea Mixon	 Randy Molina	 Aaron Molnar	 Aaron A. Montes

LaShun Moore	Tina L. Moore	Olivia Moore-Wraa	Servando Morales
Mary B. Morehouse	Mark Moreno	Nina Morris	Troye Mowrer
Rahman Muhammad	Robert A. Navarro	Justin A. Nero	Mathew Nichelini
Bruce P. Nielsen	John T. Nobriga	Kevin Nuuhiwa	Laurel Nymann

Christopher P. O'Brien	Brian J. Oftedal	Eileen Ogata	Seth P. Olyer
Enrique Orduña	Manly Ormsby	Christopher M. O'Rourke	Michael D. Osanna
Mitchell S. Ow	Celestina Pacheco	Steven Padgett	Enrique Padilla
Gerardo M. Padilla	Matt Pagsolingan	Roy Palatino	Jackson Palmer

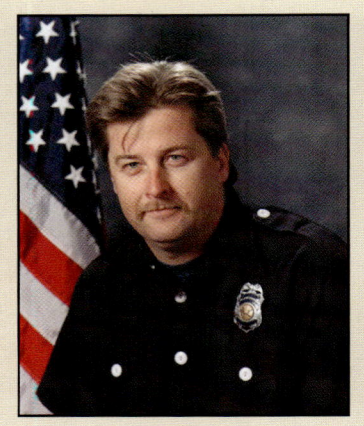

Timothy J. Pandorf	Damien D. Paraskevopoulos	Scott R. Pastor	Eric Payne
Robin Payne	Tony Perez	Christopher Peters	Tim P. Petersen
Tyehimba T. Peyton	Bradley B. Pieraldi	Preston R. Pleasants Jr.	Christie Porteous
Justin Pounds	Peter A. Pratt	Cedric Price	Roger Prieto

73

Angelo L. Primas	 Leon B. Primas	 Jerry R. Prola	 Aaron Quinn
 Mickey Quinn	 Adam Raabe	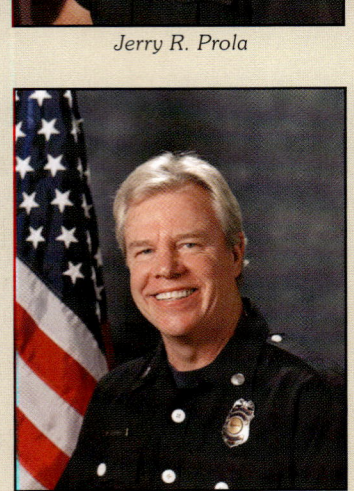  Donald C. Raabe	 Stephen M. Radulovich
Dennis J. Rainero	Jonathan D. Ramey	  Jennifer S. Ray	 James C. Ready III
 Lamar B. Redding Jr.	 Kevin Reed	 Thomas W. Reese	 John A. Reinthaler

Mark F. Reinthaler	 William A. Revilla	 Robert D. Rich	 John Richardson
 Vincent Risper	 Gary R. Robbins	 David J. Roberts	 Daniel C. Robertson
 Edmund A. Robinson	 Ernest A. Robinson III	 Leforis Robinson	 John P. Roemer
 Jose Rojas	 Larry M. Rosso	 Nathan D. Rubow	 Maria Cecilia Sabatini

Solomon C. Sacay Jr.	 Christopher S. Salas	 Nicholas X. Salgado	 Mikal A. Samad
 Larry V. Sampson	 Anthony Sanders	 Neilson Sanders	 Terrence Sanders
 Veronica Sanders	 Nicholas W. Sanz	 Daniel Sarna	 Jason G. Sautel
 Jennifer F. Schmid	 Randyn Schmidt	 Diane Schnapp	 Michael Schorr

Gary Schroeder	 Michael Scott	 Joseph C. Sermeno	 Adrian B. Sheppard
 Ernest Sherman	 Mel Shima	 Kimberly Shunk	 Joel Sibley
 German A. Sierra	 Patricia Sierra	 Demond Simmons	 Gerald A. Simon
 James Slone	 Dorothy Smith	 Harry L. Smith	 James E. Smith

James R. Smith	 Javan M. Smith	  Terence H. Smith	 Tony Soares
Damon R. Soo	 Ceasar Souza	 David H. Sparks	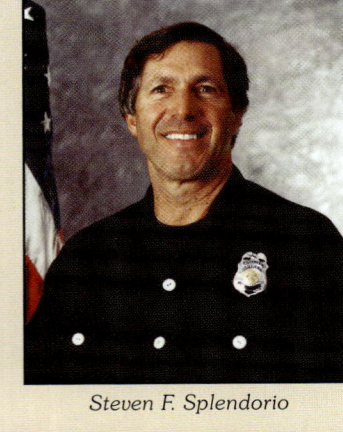 Steven F. Splendorio
 Alan A. Splithoff	  Kathy St. Thomas	 Shawn R. Stark	 Gerald L. Stein
 Jason K. Stein	 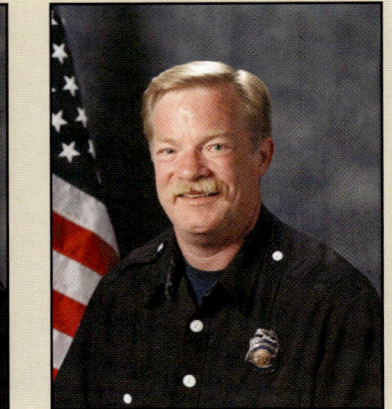 Joseph E. Steiner	 George L. Stephens Jr.	 James Stewart

Dennis Stock	Kruger W. Story	Michael E. Sullivan	Dave Swan
Pamela Y. Swan	Wayne S. Takahashi	Timothy Takis	Dai Thach
Walter Thigpen	Curtis Thompson	Sonny Thompson	Robert L. Thrower
Frank Tijiboy	Paul Tilton	Bao Q. Tô	Dino R. Torres

Joseph E. Torres	  Tim J. Tottle	 William F. Towner	 Leonard J. Townsend
 Tuan A. Tran	 William J. Triggas	 James P. Troy	 Michael Troy
 Raymond Tsang	 Clark Tucker	 Solomon Tucker	 Zachary Unger
 Emon Jamal Usher	 Jose E. Vega	 Thomas C. Veirs	 Antonio Villalobos

Charles D. Walker	Russ J. Warne	Sean Warren	Perry J. Washington
Emmanuel Watson	Ronald D. Weatherly	John M. Weir	George B. Wells
Scott West	Darin M. White	Ralph E. White	Raymond A. White
Charles Whitmarsh	James D. Whitty	Greg Williams	James Williams

Jolene Williams	 Kevin Williams	 Ulysses F. Williams	 William C. Wittmer
 Joshua Wojtkiewicz	 Linda Wong	 Terry Woodard	 Michael Worthington
 Alan N. Wraa	 Harriet Wright	 Ricky Wright	 Eva Wu
 Dennis-Ray Wylie	 David Ybarra	 David Yi	 Lawrence Young

Robert S. Young

Rodney B. Zeisse

The late Eya Yellin

Past Editor of the Local 55 newspaper and honorary member of the Bay Area Fire Forum. She was forever grateful to the fire service and the Brave New York fireman who rescued her from a fire. They were both burned severely and yet survived. Eya loved us all.

Old 3 Engine

1 Engine at City Hall.

Station 1

Station 2

Station 3

85

Station 4

Station 5

Station 6

Station 7

Station 8

Station 10

87

Station 12

Station 13

Station 15

88

Station 16

Station 17

Station 18

89

Station 19

Station 20

Station 21

Station 22

Station 23

Station 24

91

Station 25

Station 26

Station 27

Station 28

Station 29

First OFD Aircraft Rescue Fire Fighting "ARFF" Station 22 Doublewide 500 sq. ft. mobile modular trailer next to Airport Operations M-104. Opened for duty January 17, 1998 Oakland Airport.

Station 26 delivers food to Castlemont High School afterschool mentoring program.

Station 26 presents computer and printer to Castlemont High School afterschool mentoring program.

Alisa Ann Ruch Christmas party at Burn Unit of Oakland's Children's Hospital.

Station 4 delivers Christmas presents to children who wrote letters to Santa.

Station 4 tosses 60 playground balls to children at Franklin Elementary School.

Franklin Elementary School children thank Station 4 for the balls with valentines and cupcakes.

94

Two Citizen Heroes rescued a woman and her baby from an overturned car in a creek.
Inset: Random Acts Citizen Hero plaques.

Left: Citizen of the Year, Ma Green accepts Random Acts award made by Al Wraa.

Right: No Egos, Badges or Resumes Award recipient, Donna King with MC Melvin Andrews.

Captains Promotion Ceremony, 2000.

Station 20 Annual Ski Trip

Firefighters Winter Olympics

Station 15 Rafting Trip

12 Engine in the moonlight.

12 Engine hose wagon.

2544 and Station 4.

2550

2 Truck

4 Truck

2 on fire escape.

10 Engine shops.

7 Truck

4 Truck dash.

6 Engine

5 Engine crew.

EN Arellanes, BC Dossa

Another mirror bites the dust.

FF Beckley takes pride.

Captain Clean

7 Truck ladders, Station 20.

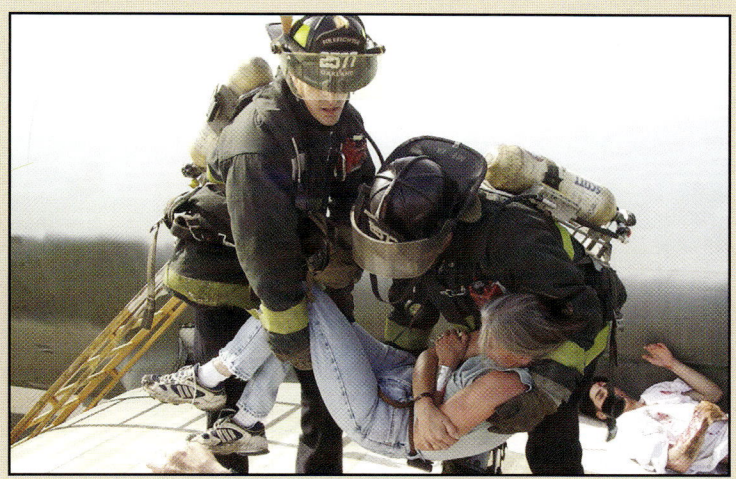

Airport drill.

Another hose bites the dust.

Argus lands at the Drill Tower.

99

Betsy Kimmel

FF Revilla sets up the aerial.

FF's Bonifacio and Blueford set the jacks.

EN Bob Rich

FF Smith and EN Bala

Brush wagon.

Change over.

Bryan Brumfield

Demond Simmons

Derek and his saw.

Delmont Waqia teaches ropes.

Dispatch Supervisor Fountaine

Crash rigs.

City Hall

Brad Pieraldi joins the hair club.

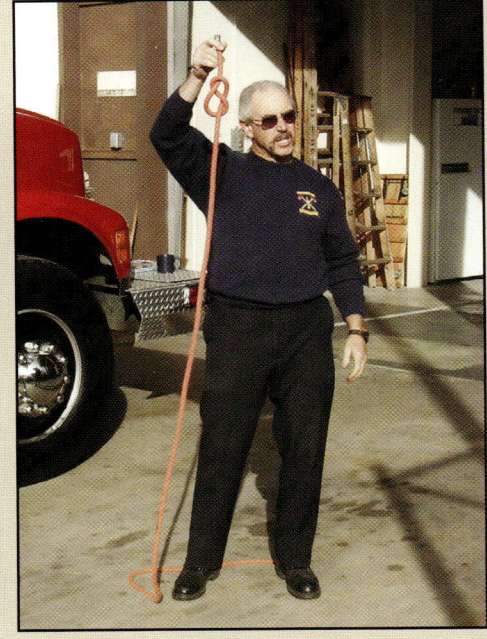

Jeff Hillstrom says this is a rope.

Howard cleans up.

Demond, Howard & Rodney

FF's Feigel Perez, and Fontalera

David Chew

Damon Soo

Cooking for burn camp.

Come and get it.

Making friends.

Mark Meiers

Jeanne Andrews

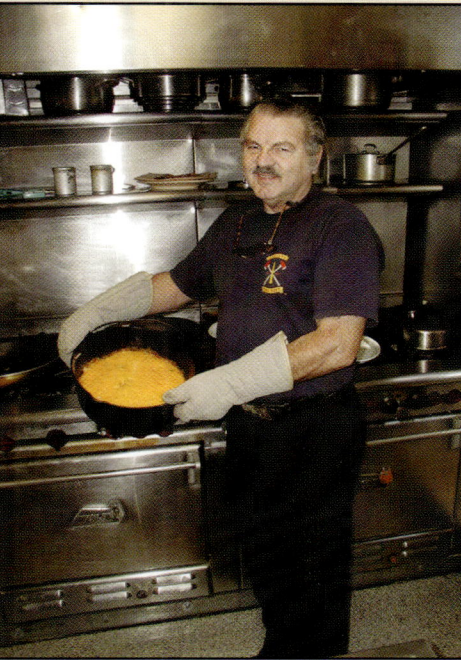

Jimmy Furtado

Paul Fellows, Geoffrey Hunter, Jason Greengrass.

103

C side of Broadway fire. (1999)

Victor Dizon and Solomon Sacay

Fenton's aftermath.

Synchronized overhaul.

Leon Primas

Kevin Kennedy

Study, study, study.

Willie Dixon, Michael Sullivan, and James Reis

Doug Abbott

Ray Dossa and John Ironside

Downtown

Drill till it's right.

Downtown Oakland

Engine 5

105

Station 20 roll call.

Entry Class

David Espino & Damon Covington practice EMS.

Family Hour at 18.

Here Kitty.

Five Truck Crew.

Green spare.

Harvell Guiton

Ed Robinson and friends.

Ian McWhorter at the Tower.

I can make anything out of wood.

Drill Tower Clouds.

107

I love to cook.

I'll be up here.

Jay Comella and his smooth bore.

Group photo.

Coach

At the foot.

Howard takes his hygiene seriously.

If you think I'm going to retire just so you can get a badge …

"For Recreation Only"

If a little CO_2 is good, a lot is better.

It just doesn't get any better.

Infection control.

I caught one!

Ceasar Salad

J.P. Troy — Big Man on scene.

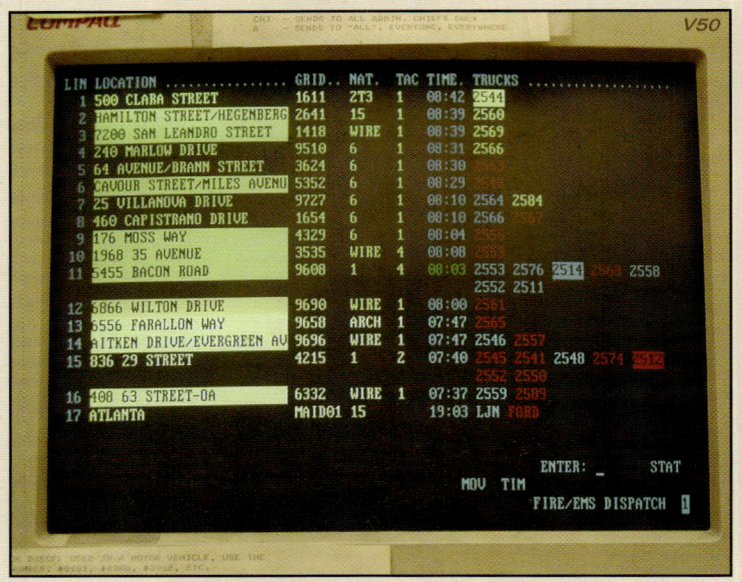

Umm … we're kinda busy right now.

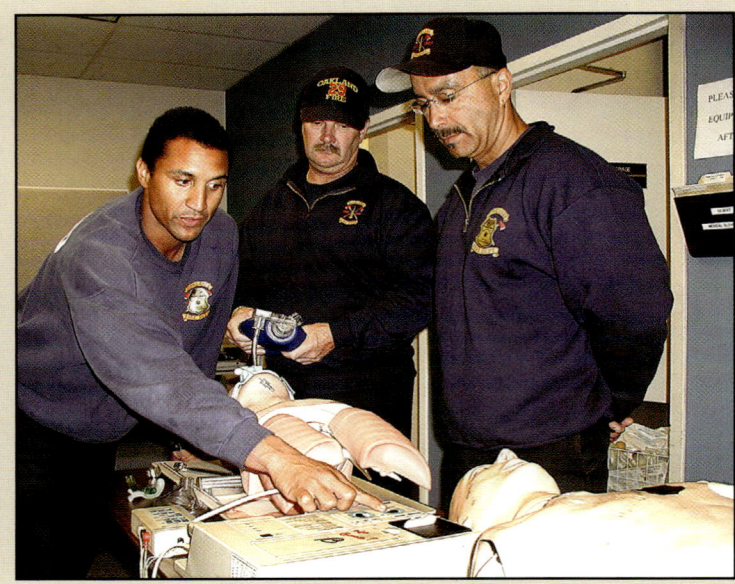

Liggins, Beede and Ybarra.

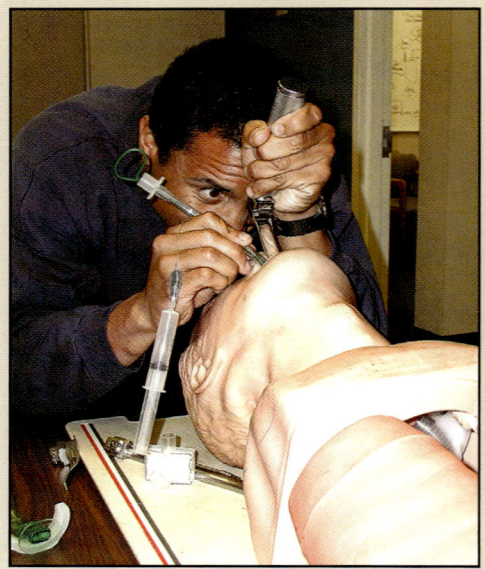

Liggins

John Kuehl, Larry Young, and Greg Bell.

Larry Allworth

110

Lloyd White

Look Up

Mark Maddox says "Hi."

Max was the best.

Angelo Primas teaches ladderpipe.

111

Chuck Accurso

Frank Wong - Photo 1

Ladder Pipe Drill.

18th and Lakeshore.

Ventura Firefighter's Olympics

Softball Team

Sta. 8 B-Shifters

Medina can cook.

Meineke, Calhoun and Flashberger.

Now that's a roof.

Now that's a fire.

Mike Donner likes to teach.

O'Brien and Beckley paint axes.

113

New kids.

OFD in the heat.

Meineke bagpipe.

FF's Moore and Llamas

Michael Asport

Wait for orders.

Paul Fellows

Under the hood.

114

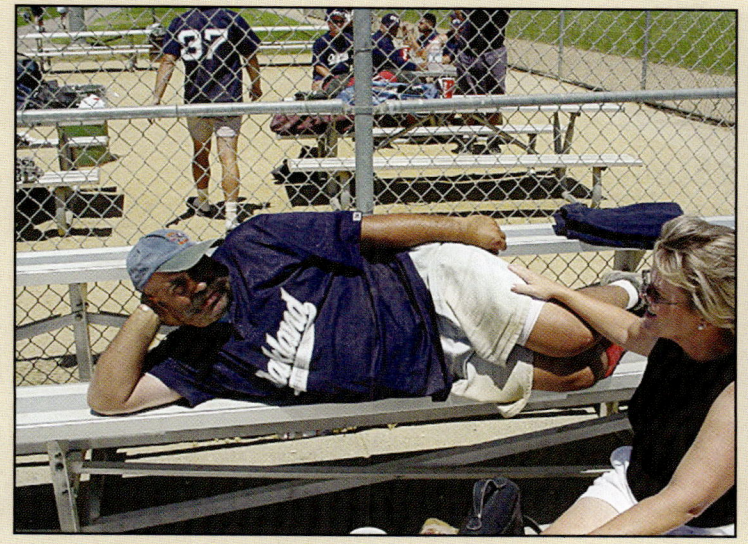

Brad's boudoir pose.

John Farrell, bulking up.

FF Feirerra, EN Farrell, FF Garcia

Union leadership, hard at work.

Paul Fellows puts it away.

115

Brad Pieraldi ... and twin?

Kenny and Gina

Ken Cuneo

The new Station 8.

Ladder pipe.

Sean and Doug.

Reinthaler, Walker

Pablo Cruz

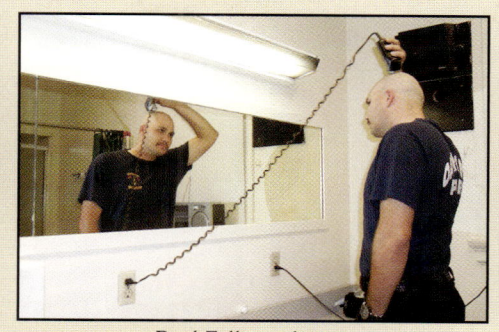

Paul Fellows shaves.

AARBF Burn Relay

Pickle ball is our life.

Ralph Kurio keeps a cleaning.

Redding, Dizon, and Chew.

Redding, Takahashi and Lopes.

Ready to drill.

Pointy part.

Rick and Chip know overhaul.

117

Rick and Bruce try on red hats.

Rick keeps an eye out.

Propane drill.

PJ tends the grill.

Ralph Kurio

Schmid, Martinez and Liggins.

Rusty Keyes hauls bottles.

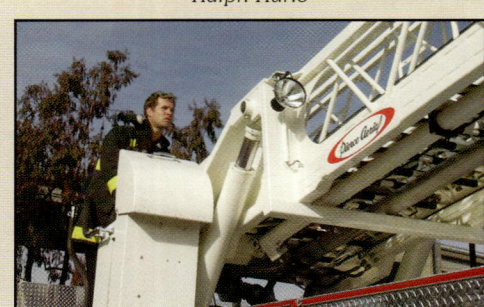

Rodney Zeisse

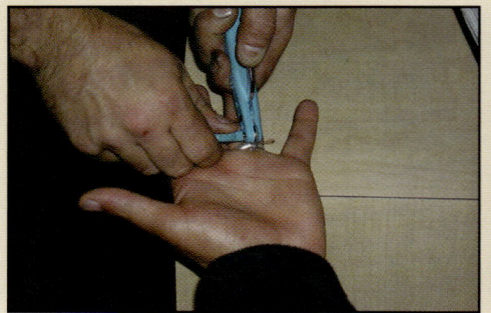

Ring Cutter

Shawn Stark

Shop 2.

Dennis Holmes and a young ride-along.

Robbins and Heath on the roof.

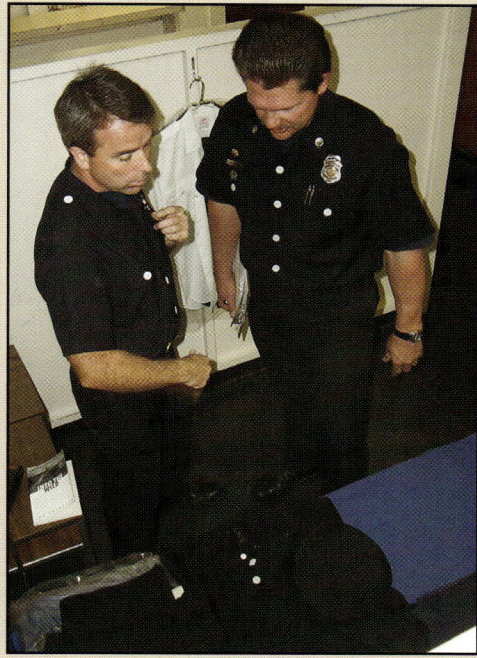

Roberts and McKee.

Ralph Kurio and his engine.

Randy Medina

Rick Young is the greatest.

Stephanie Massey

Stettler, Drennan, D. Chew.

Steaks On!

119

Rick makes a friend.

Soares and a friend.

Russ Earle makes cookies.

Stark, Gascie

Set the dogs.

Station 10's watch board.

Rank and file.

Terrence Smith

The coach shows how it's done.

120

Saluting our New York brothers.

Tommy is prepared for anything.

Too little, too late.

Eldon Parker at an auto extrication.

Tony Crudele

Tony Jackson

Villalobos on the monitor.

121

Willie Dixon

Tim Takis hustles.

Turnout coats.

Somebody loves us.

Valida Holmes

USAR drill.

Water

Which one is better?

Triggas and a friend.

What do these two men have in common?

What do you think Alex?

Young can cook.

122

Tim's last day at Station 4.

Lyle and Phil enjoying another dinner.

Rick Lapora, Ken Costa, Al Wraa, Harry Van Arsdale and Stick Morgan celebrating another 4 E Retirement.

Food fight Station 4.

"A Few Good Men." Steve Miguel, Charles Walker, Harry VanArsdale, Rick Lapora, Tim Petersen, Dan Robertson, George Freelen, Steve Radulovich.

Mark Fraser, Gary Robbins, and Steve Splendorio.

Tannehill's last shift. Standing: Dave Roberts, Joe Rudiger, Rich DeGlymes. Seated: Lee Tannehill and John Buckhorn.

Classmates — Where is the Bean? From left: Splendorio, Tannehill, DeGlymes, Wraa, Station 4.

Lee Tannehill final shift. Jerome Hathaway and Lee Tannehill.

1996, 4 Eng. 2 Trk "B" Shift. Front row: Dwight "Bean" Ontiveros. Lyle Elvin, Ken Costa, Jack Stein. Back row: John Ferguson, Jay Comella, Al Wraa, John Larson, Lee Tannehill.

Mike Hoskins, Dan Robertson, Tony Di Stefano.

Al Edwards, 4 Eng 2 Truck, Tinker — who?

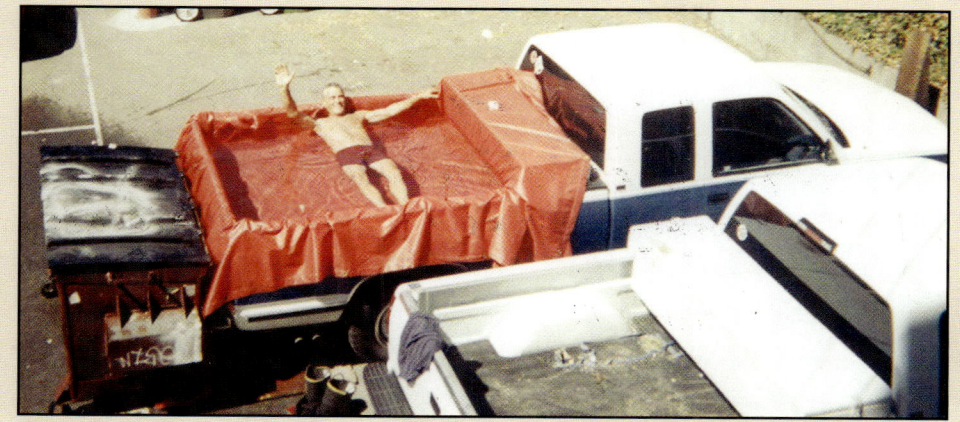

"Coolin' off after a rough fire!" 4 Eng 2 Trk.

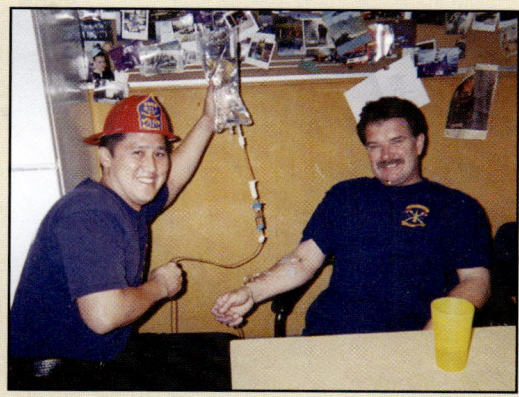

"A & O times zero!" Station 23, FF Keith Hall, FF Ben Beede, December 2001.

Front row: Jerome Hathaway, Ken Costa, Lyle Elvin, Dwight "Bean" Ontiveros. Back row: Steve Miguel, Jay Comella, Lee Tannehill, Al Wraa, John Larson, and John Ferguson.

Station 28, A Shift. From left: Eng. Mike Osanna, FF Jim Dillon, Lt. John Connelly, FF Prestin Baker.

Cigar Time at Station 23.

Station 23 receives a trophy for annual inspection. Capt. Tom Gallinatti and Chief Simon, February 2002.

125

Zach Hilton poses for a picture after a house fire, May 22, 2002. He is also known as "Charlie Brown."

Brian and Zach in front of Engine 27.

Firefighter Chris Colgan on an Emeryville fire, April 2001.

Lorenzo Ellison after a house fire, May 22, 2002.

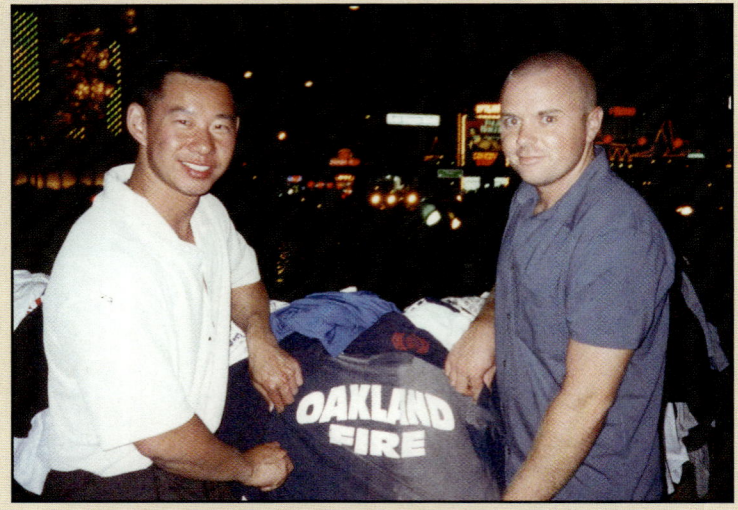

Leonard and Zach show off OFD shirt placed by them in Las Vegas. Never Forget.

Zach, Shad, Javan. Bike paramedics for Sept. 11th remembrance event.

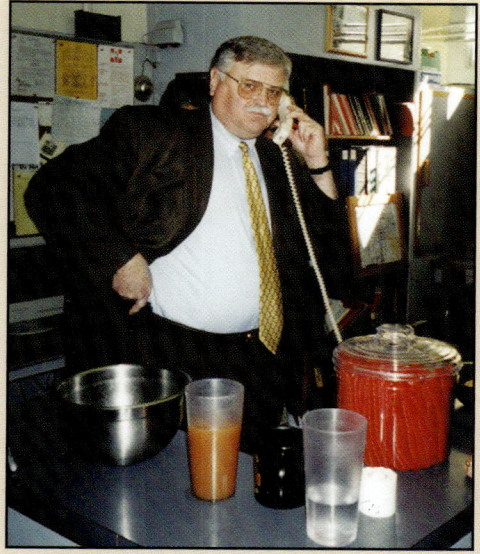

Lt. Ed Clausen (Ret.) taking care of business.

Station 6 "A" shift. From left: F/F Tom Reese - EN. Steve Splendorio, LT. Abe Merritt, Firefighter Paramedic Bud Souza … Bud, Bud, where is Bud?

Shad Jessel in his uniform for the first bike paramedic event. Javan Smith is his partner for the day.

Al Grasso, Jack Stein, Steve Baptista and Jim Colussi, Station 4.

German Sierra making sure he has enough PCRs for his first bike paramedic patrol.

Old guys get together, "Free Geritol Giveaway."

16 E with 2 Truck after a site inspection, Station 4B.

Captain Jarrett congratulating the "New Captain Pandorf."

Long lineage of engineers from 4 Eng 2 Trk. From left: Lapora, Petersen, Fraser, Fitzgerald and Wraa.

Three retired chiefs: Lyle Elvin, Jim Colussi, Don Matthews.

4 Eng 2 Trk B Shift Brothers.

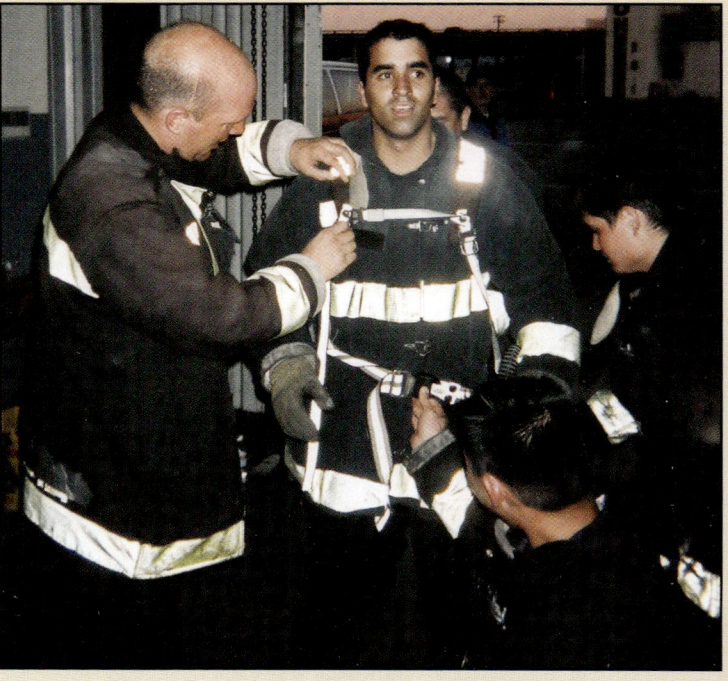

Kenny Ergun during the academy. November 2000.

From left: FF Brad Pieraldi (5 Trk), FF Ron Martinez (5 Trk), FF Robert Beckley (5 Trk), FF Seth Olyer (5 Trk), FF Bao Tô (6 Trk), LT. Lawrence Hom (8 Eng.), FF Mike Donner (5 Trk), FF Roger Prieto (8 Eng), EN Phil Garcia (19 Eng). At the top of the Empire State Building, after attending the funeral service of a fallen brother of the FDNY, Nov. 10, 2001.

From left: Frank Tijiboy, Perry Magill, Paul Fellows, Lorenzo Frediani, 6 Truck, 2001.

Class 2-00, during the recruit adacemy. The cake that Dino wouldn't eat! November 2000.

Strike Team to Lake Shasta, CA, 2000. Ralph Kurio, Robert Lipp and Paul Fellows.

FF's Hilton, Oftedal, Smith after graduation, December 12, 2000.

From left: Hilton, Oftedal, Torres, Wittmer, Smith and Nadia. Graduation class of 2-00, December 2000.

129

Hilton and Waqia on a car fire, February 2001.

Old guys at retirement dinner.

Frank Tijiboy, Perry Magill on roof, 6 Truck 2000.

Lee's last supper.

130

Recruit class 2-00 with Chief Simon, December 2000.

Javan Smith finishes star drill at the Academy with help from (left to right) Michael Hall, Damien Paraskevopoulos and Kenny Ergun.

From left, front row: Mark Fraticelli, Ken Cuneo, Paul Leimone, Mark Meiers, Tony Distefano, John Reinthaler. Back row: Brad Pieraldi, Jimmy Smith, Steve Miguel, Glenn Gonzalez, Dan Robertson, Kevin Reed, Lynn Bauman. January 1995 ski trip.

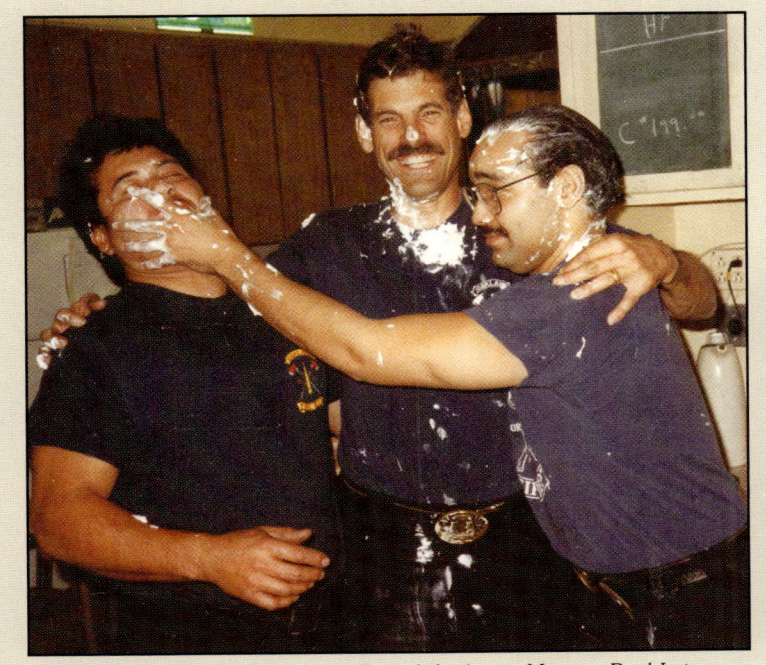

Playing around at old 20 E 93/A. From left: Aaron Montes, Paul Leimone, Damian Bala, 1990s.

Cooling off in "home-made" cold tub at old E 20. From left: Joe Hoke, Damian Bala, Mike Donner, 1990.

The Wraas.

Carlos Harvey, August 1999.

4 Engine and 2 Truck "B" Shift.

8 Trk at 23 Engine, 1976.

Union Installation.

Pat Stranahan, Special Olympics.

Old 3 Engine, Bob Speas and Mike Osanna.

John Connelly and Jerry Speka at Union Picnic.

Hoseman Harry VanArsdale tillering 1T in 1973.

Old Station 3, Haslip (Marty) Weber and Walt Fredrickson with camera.

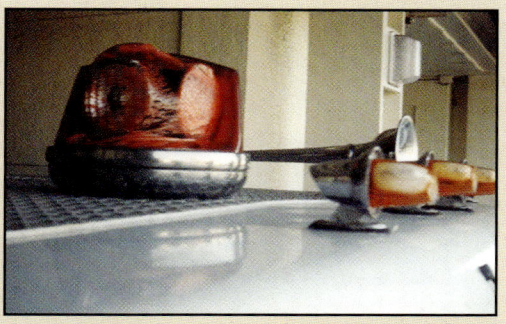

6 Trk melted beacon from fire at 82nd Ave. and San Leandro St.

Station 4 B shift, 1989. Ron Weatherly's last day as a firefighter.

Randy Kirchner and Eugene Dick putting patch kit on.

Old 3 Engine. From left: Howard Strawn, Dennis Holmes, Walt Klingler, Don Parker, Bob Comella, John Connelly, Joe Rudiger, Marty Rinne, Mark Hoffmann, AC Bob McGue.

Hoseman Harry VanArsdale tillering 2 Trk, 25 years later.

133

As part of OFD ARFF Airport Community Service station 22 ARFF apparatus performs a traditional "Water Bridge" for inaugural flights, retirements, and other significant events for Oakland International Airport.

2,000 sq. ft. apparatus airport Fire Station #22 M-9-11.

John (F.B.) Buckhorn and Everett "Stick" Morgan, 4 Eng. 2 Truck, "Retired."

Marshall McKee, Robert "Bear" Beckley, Dave Roberts and Ed Schennek.

From right: Engineer Kathy Campbell, Firefighters/Paramedics Damien Paraskevopoulos and Annette Goodfriend.

Oakland Fire Department Index

— A —
Abbott, Doug 51, 105
Abel, Susan L. 51
Abram, Douglas S. 51
Accurso, Chuck 51, 112
Adams, Jake 51
Adams, Lenroy 51
Aff, Edward M. 51
Aguilar, Richard 51
Allworth, Larry 110
Amormino, Matt B. 51
Anderson, Paul 51
Andrews, Jeanne 36, 51, 103
Andrews, Melvin 51, 95
Andrion, Nerissa 51
Antes, Dominic R. 51
Archuleta, Alex E. 51
Arellanes Jr., Cruz 51, 98
Ashley, Darryl J. 52
Askew, Lisa J. 52
Asport, John M. 52
Asport, Michael 114
Austin-Garrett, Joan 43, 52

— B —
Baker, Charles A. 52
Baker, John K. 39
Baker, Lisa A. 52
Baker, Prestin L. 125
Bala, Damian 52, 131
Bala, Joanelle 52
Ball, Nicholas A. 16, 39
Banks, Raymond V. 52
Banks, Sherri 52
Baptista, Steve 127
Barlogio, Danny 2
Barr, Daniel J. 18
Basada, Phillip 52
Bauman, Lynn 131
Beckley, Robert 52, 99, 113, 129, 134
Becton, Lamont B. 52
Beede, Benjamin L. 52, 110, 125
Beldiman, Gabriela B. 52
Bell, Coleen Bell 52
Bell, Gregory D. 53, 110
Bell, Phillip R. 53
Bernard, Tanisha B. 53
Berry, Erick D. 53
Berry, Glenn D. 41, 53
Bird, Roger E. 53
Blueford II, Jerry E. 43, 53, 100
Bobb, Robert C. 5
Bolin-Chew, Eleanor M. 53
Bolton, Odessa 43
Bonifacio, Neptali B. 53, 100
Bowers, Kellwin L. 53
Bowron, James 53
Bozman Jr., Donald R. 53
Bragg, Ronald A. 41
Brandle, Marlon 53
Brierly, John Ryan 53
Briscoe, Daphne 53
Brown, Jerry 6
Brown, Sean 53
Brownlow, Nathaniel 54
Brue, David A. 54
Brumfield, Bryan 54, 101
Brunson, Senator 42, 54
Buckhorn, John 124, 134
Budiao, Paul B. 42, 54
Buell, Linda 54
Burguemo, Barbara 54
Burke, James H. 26, 39
Burns, Betty 54

— C —
Calhoun, James 54, 113
Campbell, Kathleen 54, 134
Campos, Ron 33
Carlton, George H. 14
Carlyon, John 29, 49
Carter, Ronald K. 54
Carter, Steven D. 54
Case, Gregory J. 54
Castillo, Mario 54
Catalano, James R. 54
Catano, Kim 43, 44, 54
Caulfield, Joe 55
Cazenave, Lucien B. 55
Cetraro, Louis 49
Chan, Denny 55
Chaney-Williamson, Sylvia 44, 55
Chapital, André B. 55
Chew, David E. 55, 102, 119
Chew, Richard 43, 55, 119
Chew, Rick 2, 86, 117
Chew, Steven 55
Chiguila, Gilberto 55
Chin, Gene 55
Chin, Tracey Ann Rene 55
Clark, Chris 43, 44, 55
Clausen, Ed 38, 127
Clean, Capt. 99
Clemons, Weldon 55
Cody, Gilbert M. 43, 55
Coffer, John A. 55
Colgan, Christopher 55, 126
Collins, Gary C. 44, 56
Colussi, Jim 127, 128
Comeaux, Julian 56
Comella, Bob 133
Comella, Jay 56, 108, 124, 125
Comstock, Jesse R. 56
Concepcion, Ritalinda 56
Conley, Gary K. 56
Conley, James 56
Connelly, John 56, 125, 132, 133
Costa, Kenneth M. 56, 123, 124, 125
Covington, Damon 56, 106
Cox, Bonnie 56
Crudele, Tony 33, 121
Crum, Clarence 38
Cruz, Pablo 56, 117
Cuevas, Victor M. 56
Culberson, Carolyn 56
Cuneo, Kenneth 56, 116, 131
Curtis, Jacqui 56

— D —
Dalrymple, Chris B. 57
Danziger, Steve 57
Dawkins, Heather 57
De Glymes, Rich 57
de la Montanya, Matthew 12, 39
De Lacy, James R. 57
De Silva, Elaine 57
DeGlymes, Rich 124
Delgado, Kenneth 57
DeMarco, Floyd 29, 49
Detlefsen, Hanns 57
Detlefsen, William D. 57
Di Stefano, Anthony C. 57, 124, 131
Dick, Eugene 133
Dillon, Jim 125
Dixon, Willie 57, 105, 122
Dizon, Victor 57, 104, 117
Dolan, John M. 57
Domingo, Renee 57
Domino 43
Donaldson, Tracy 57
Donner, Michael D. 57, 113, 129, 131

Doody, Miles 39
Dossa, Ray 98, 105
Douglas, Mark C. 58
Douglas, William B. 58
Draper, Eric 58
Drayton, Melinda Ann 58
Drennan 119
Dunham, Glenn 58
Duvall, James W. 58
Dwyer, Daniel T. 58

— E —
Eade, Timothy H. 58
Earle, Russell J. 58, 120
Edwards, Al 124
Edwards, James D. 58
Eger, David 58
Elento, Thomas 58
Elliott, Justin 58
Elliott, William 31, 49
Ellison, Lorenzo 58, 126
Elvin, Lyle 124, 125, 128
Emerson, Jackie 58
Emke, Kurt 58
Epps, Harold Ray 59
Ergun, Kenny 59, 128, 131
Erhardt, Kenneth B. 59
Espino, David 59, 106
Evans, Keith A. 59
Ewell, P. Lamont 39

— F —
Fahey, Emmett J. 38, 59
Fair, William H. 39
Fairley-Summers, Helen 59
Farrell, Daniel D. 39, 59
Farrell, John 59, 115
Feigel-Perez, Richard 59
Feirerra 115
Fellows, Paul 7, 34, 35, 37, 38, 59, 103, 114, 115, 117, 129
Felton, J. B. 12
Ferguson, John M. 59, 124, 125
Ferreira, Mark D. 59, 115
Ferrer, Moses 59
Fesai, Sergei 59
Fincher, Keenan 59
Fire Safety and Education Day (Week) 43
Firefighters Winter Olympics 96
Fires of the 1860s 11
Fires of the 1870s 14
Fires of the 1880s 15
Fires of the 1890s 16
Fires of the 1900s 18
Fires of the 1910s 19
Fires of the 1920s 23
Fires of the 1930s 23
Fires of the 1940s 26
Fires of the 1950s 27
Fires of the 1960s 29
Fires of the 1970s 30
Fires of the 1980s 32
Fires of the 1990s 35
Fitzgerald 128
Flashberger, Keith E. 60, 113
Foley, Christopher A. 60
Fontelera, Bruce 59, 102
Ford, Debbie 60
Foutaine, Annette 60
Fraker, Guy 43
Fraser, Mark 36, 123, 128
Fraser, Zachary W. 60
Fraticelli, Mark A. 60, 131
Frediani, Lorenzo 60, 129
Fredrickson, Walt 133
Freelen, George T. 60, 123
Fujii, Charlotte 60

Fuller, Fred O. 39
Furtado, Jimmy 103

— G —
Gadison, Harold N. 60
Gadison, Neil 41
Gallinatti, Thomas D. 60, 125
Garcia 115
Garcia, Charles A. 60
Garcia, Dwight 60
Garcia, Philip 60, 129
Garcia, Robert 60
Gardea, Monty J. 61
Gardner, Carl 7, 61
Gascie, Sean Edward 61, 120
Gaskin, Darryl N. 61
Gaul, Alan 61
Gentry, Neil 61
Georgatos, Nicko 61
Glass, Charles 61
Golden, Samuel 32, 39
Gomez, Hernán 61
Gonsolin, Scott 61
Gonzalez, Bobby 61
Gonzalez, Glenn S. 61, 131
Goodfriend, Annette 34, 61, 134
Grasso, Al 127
Gray, Dwayne 61
Green, Julie E. 41, 61
Green, Kim M. 61
Green, Ma 95
Greengrass, Jason A. 62, 103
Gregoire, George G. 62
Gregoire, Roland R. 62
Gresher, Charles 62
Griffin, Leroy 62
Griggs Jr., Ronald E. 62
Grisby, Sylvester 62
Gross. James A. 62
Grunwaldt, Carlos A. 62
Guiton, Harvell 62, 107
Gullett, Gordon F. 62

— H —
Hajny, Derek 62
Hall, Keith 38
Hall, Keith Michael 37, 62, 125
Hall, Michael D. 62, 131
Halley, J. C. 11
Halliday, David K. 62
Halpin, James 62
Hammond, Connie 63
Harger, Bradley L. 63
Harris, Donnie J. 63
Harris, Gregory J. 63
Harris, James 41
Harris, Yvette 63
Harvey, Carlos 7, 63, 132
Hathaway, Jerome 63, 124, 125
Haughy II, James Patrick 63
Haughy, Jeffrey 63
Hayes, Daniel D. 13
Heath, Chris R. 63, 119
Hector, David E. 63
Heelan, Thomas 49
Heinrich, Christen 63
Hellige, Scott G. 63
Hendricks, Larry 45, 63
Hill, James 13, 39
Hill, Michael 63
Hill, Mike 41, 42
Hillesheim, Michael 63
Hillstrom, Jeff 102
Hilton, Zachary C. 64, 126, 129, 130
Hines, David 64

Hoard, Howard 64
Hoffmann, Mark 64, 133
Hogerheide, Derek A. 64
Hoke, Joseph 64, 131
Holmes, Dennis 119, 133
Holmes, Jacob A. 64
Holmes, Valida D. 64, 122
Holt, Howard C. 64
Hom, Donna 64
Hom, Lawrence S. 64, 129
Honery, Bryan 33
Honeycutt, Neil 38
Hoskins, Michael A. 64, 124
Hoskins, Mikhail 64
Howard, Richardine 64
Howell, Jim 43, 44
Hunter, Geoffrey 38, 64, 103

— I —
Ibarra, Eduardo H. 64
Idle, Paul T. 41, 65
Ironside 105
Irwin, John 65

— J —
Jackson, Anthony R. 65, 121
Jackson Jr., Johney L. 65
Jackson, Wellington 65
Jain, Vibhor 65
James, Larry 65
James, Richard 65
Jarrett, William 65, 128
Jeddi, Porya 65
Jensen, Ann Marie 65
Jessel, Shad D. 65, 127
Johnson, Ronald 65
Jones, Keith L. 65
Jones, Sherman 65
Joseph, James A. 65
Jung, Leonard T. 66
Justice, Coy M. 66

— K —
Kafele, Malik H. 66
Kaigler-Armstead, Margaret 66
Katsanos, Lee D. 66
Keenan, Dan 38, 66
Keeton, Mary 66
Kennedy, James F. 14, 16, 39
Kennedy, Kevin M. 66, 104
Keyes, Rusty 66, 118
Killingsworth, Fitzroy E. 66
Killingsworth, Stanley E. 42, 66
Kilmartin, Edward J. 66
Kimmel, Betsy 66, 100
King, Donna 66, 95
Kirchner, Randy E. 32, 66, 133
Kittel-Nwuke, Denise 66
Klingler, Gary C. 67
Klingler, Walt 133
Knaus, Mathew 67
Knudsen, Jamie M. 67
Kouhi, Arosh 67
Kozak, Rebecca 67
Kuehl, John D. 67, 110
Kurio, Ralph W. 67, 117, 118, 119, 129

— L —
Laffen, Sean D. 67
Landeza, Father Jayson 67
Landry, Christopher 67
Langford, Dwight 67
Lapora, Richard C. 67, 128
Lapora, Rick 123
Larson, John 36, 67, 124, 125

135

Lawton, Elburton B. 39
Layman, Lloyd 23
Leary, Shonda 67
Legear, Dennis 67
Leimone, Paul J. 33, 67, 131
Lendl, Carl 68
Leslie, Frank R. 68
Liggins, Daryl 68, 110, 118
Lipp, Robert 68, 129
Lizama, Alvaro 68
Llamas, Edward 68
Llamas 114
Lloyd, Jason A. 68
Logan, Erik N. 68
Logan, Lisa 68
Lopes, Brian J. 68, 117
Luby, Nickolas 68
Lutkey, William G. 22, 26, 39

— M —
Macor, Michael 36
Macpherson, David 68
Maddox, Mark 68, 111
Magill, Perry B. 68, 129, 130
Mahoney Jr., Lester 68
Marrero, Santalynda 68
Marshall, Dan 69
Marshall, Nola 69
Marshall, Thomas 69
Martin, Janet 69
Martinez 118
Martinez, Jeromie 69
Martinez, Peter G. 41, 69
Martinez, Ronald 69, 129
Martinez, Walter 69
Massey, Stephanie 43, 119
Mathis, Charles 69
Matthews, Don 128
Matthews, Keith 69
McCarthy, Mike 44
McCarty, James A. 69
McCrary, Scott 69
McFeeley, J. E. 19
McGue, Bob 133
McGuire, Thomas 11
McKee, Marshall S. 38, 69, 119, 134
McNicholas, James 69
McWhorter, Ian 69, 107
Medina, Randy L. 69, 113, 119
Meiers, Mark 70, 103, 131
Meineke, Ryan A. 70, 114
Mejia, Manuel 70
Menge, Dennis R. 70
Menietti, Stephen L. 39
Mercado, Christian 70
Merritt, Abe 127
Merritt, Ahvrumm 70
Merritt, Samuel 15
Miguel, S. 36
Miguel, Stephen 70, 123, 125, 131
Miles, Roy A. 70
Miller, Earl Lee 70
Miller, June 79
Miller, Michael E. 70
Miranda, Maurice R. 70
Mixon, Thea 70
Moffitt, James 14, 16, 39
Molina, Randy 70
Molnar, Aaron 70
Montes, Aaron A. 36, 71, 131
Moore 114
Moore, LaShun 71
Moore, Tina L. 71
Moore, William 39
Moore-Wraa, Olivia 36, 71
Morales, Servando 71
Morehouse, Mary B. 71

Moreno, Mark 71
Morgan, Everett "Stick" 123, 134
Morris, Nina 71
Mowrer, Troye 71
Muhammad, Rahman 71

— N —
Nadia 129
Nalley, John C. 39
Navarro, Robert A. 71
Nero, Justin A. 71
Nichelini, Mathew 71
Nielsen, Bruce P. 71
Nobriga, John T. 71
Nuuhiwa, Kevin 71
Nymann, Laurel 71

— O —
Oakland Black Firefighters Association 47
O'Brien, Christopher P. 72, 113
Office Of Emergency Services (OES) 46
Oftedal, Brian J. 72, 129
Ogata, Eileen 72
Olyer, Seth P. 72, 129
Ontiveros, Dwight "Bean" 124, 125
Orduña, Enrique 72
Ormsby, Manly 72
O'Rourke, Christopher M. 72
Osanna, Michael D. 72, 125, 132
Ow, Mitchell S. 72

— P —
Pacheco, Celestina 72
Padgett, Steven 72
Padilla, Enrique G. 72
Padilla, Gerardo M. 72
Pagsolingan, Matt 72
Palatino, Roy 72
Palmer, Jackson 72
Pandorf, Timothy J. 73, 128
Paraskevopoulos, Damien D. 73, 131, 134
Parker, Don 133
Parker, Eldon 121
Parker, Frank 18, 49
Pastor, Scott R. 73
Payne, Eric 73
Payne, Robin 73
Perez, Feigel 102
Perez, Tony 72
Peters, Christopher 73
Petersen, Tim P. 73, 123, 128
Peterson, Lance 33, 49
Peyton, Tyehimba T. 73
Pieraldi, Bradley B. 73, 102, 116, 129, 131
Pierson, Walter 39
Pimentel, Joseph 49
Pleasants Jr., Preston R. 73
Porteous, Christie 73
Pounds, Justin 73
Pratt, Peter A. 73
Price, Cedric 36, 73
Prieto, Roger 38, 73, 129
Primas, Angelo L. 74, 111
Primas, Leon B. 36, 74, 104
Prola, Jerry R. 74

— Q —
Quinn, Aaron 74
Quinn, Mickey 43, 74

— R —
Raabe, Adam 74
Raabe, Donald C. 74
Radulovich, Stephen M. 74, 123
Rainero, Dennis J. 74
Ramey, Jonathan D. 74
Randon Acts of Kindness 45
Ray, Jennifer S. 74
Ready III, James C. 74
Ready, James M. 19
Records, Dewey 49
Redding Jr., Lamar B. 74, 117
Reed, Kevin 74, 131
Reese, Thomas W. 74, 127
Reid, Larry 43
Reinthaler, John A. 43, 74, 116, 131
Reinthaler, Mark F. 75
Reis, Jim 105
Revilla, William A. 75, 100
Rich, Robert D. 75, 100
Richardson, John 75
Riley, James 36, 49
Rinne, Marty 133
Risper, Vincent 75
Robbins, Gary R. 75, 119, 123
Roberts, Dave 119, 124, 134
Roberts, David J. 75
Robertson, Dan 123, 124, 131
Robertson, Daniel C. 75
Robinson, Edmund A. 75, 107
Robinson III, Ernest A. 75
Robinson, Leforis 75
Rodgers, Camille 43
Rodrigues, Sheree 37
Roemer, John P. 75
Rojas, Jose 75
Rosso, Larry M. 42, 75
Ruch, Alisa Ann 94
Rudiger, Joe 124, 133

— S —
Sabatini, Maria Cecilia 75
Sacay Jr., Solomon C. 76, 104
Salas, Christopher S. 76
Salgado, Nicholas X. 76
Samad, Mikal A. 76
Sampson, Larry V. 76
Sanders, Anthony 76
Sanders, Neilson 76
Sanders, Terrence 76
Sanders, Veronica 76
Sanz, Nicholas W. 76
Sarna, Daniel 76
Sautel, Jason G. 76
Schennek, Ed 134
Schmid, Jennifer F. 34, 76, 118
Schnapp, Diane 76
Schorr, Michael 76
Schroeder, Gary 77
Scott, John W. 11, 39
Scott, Michael 77
Sermeno, Joseph C. 77
Sheppard, Adrian B. 77
Sherman, Ernest 42, 77
Shima, Mel 77
Short, Samuel H. 21, 39
Shunk, Kimberly 77
Sibley, Joel 77
Sicotia, A. A. 18
Sierra, German A. 77, 127
Sierra, Patricia 77
Silveria, Terry 29, 49
Simmons, Demond 77, 101
Simon, Gerald A. 4, 39, 43, 77, 125, 131
Slone, James 77
Smith 100
Smith, Dorothy 77

Smith, Harry L. 77
Smith, James E. 77
Smith, James R. 78
Smith, Javan M. 78, 127, 129, 131
Smith, Jim 42
Smith, Jimmy 131
Smith, Terence H. 78, 120
Smith, W.B. 19
Soares, Herbert A. 120
Soares, Tony 78
Soo, Damon R. 78, 102
Souza, Bud 127
Souza, Ceasar 78
Sparks, David H. 78
Speas, Bob 132
Speka, Jerry 132
Splendorio, Steven F. 78, 123, 124, 127
Splithoff, Alan A. 78
St. Thomas, Kathy 78
Stark 120
Stark, Shawn R. 78, 118
Station 1 85
Station 2 85
Station 3 85
Station 4 86, 94, 123, 124
Station 5 86
Station 6 86
Station 7 87
Station 8 87, 116
Station 10 87
Station 12 88
Station 13 88
Station 15 88, 96
Station 16 89
Station 17 89
Station 18 89
Station 19 90
Station 20 90, 96, 106
Station 21 90
Station 22 911
Station 23 91, 125
Station 24 91
Station 25 92
Station 26 92, 94
Station 27 41, 92
Station 28 93
Station 29 93
Stein, Gerald L. 42, 78
Stein, Jack 124, 127
Stein, Jason K. 78
Steiner, Joseph E. 78
Stephens Jr., George L. 78
Stettler 119
Stewart, James 78
Stock, Dennis 79
Story, Kruger W. 41, 79
Stranahan, Pat 132
Strawn, Howard 133
Sullivan, Michael E. 79, 105
Swan, Dave 79
Swan, Pamela Y. 79
Sweeney, James J., Jr. 26, 39

— T —
Takahashi, Wayne S. 79, 117
Takis, Timothy 79, 122
Tannehill, Lee 124, 125
Taylor, George 11, 12, 39
Taylor, Godwin 39
Thach, Dai 79
Thigpen, Walter 79
Thompson, Curtis 79
Thompson, Sonny 79
Thrower, Robert L. 79
Tijiboy, Frank 79, 129, 130
Tilton, Paul 79
Tô, Bao Q. 79, 129
Toomey, Tracy 36, 49
Torres 129
Torres, Dino R. 79

Torres, Joseph E. 36, 80
Tottle, Tim J. 80
Towner, William F. 80
Townes, Royal 23
Townsend, Leonard J. 80
Tran, Tuan A. 80
Triggas 122
Triggas, William J. 80
Troy, James P. 80, 110
Troy, Michael 80
Tsang, Raymond 80
Tucker, Clark 80
Tucker, Solomon 80

— U —
Unger, Zachary 80
Usher, Emon Jamal 80

— V —
VanArsdale, Harry 123, 132, 133
VanGorder, Kenny 37
Vega, Jose E. 80
Veirs, Thomas C. 80
Villalobos 121
Villalobos, Antonio 80

— W —
Waalkes, Ralph 49
Walker, Charles D. 81, 116, 123
Walsh, Barney 28
Waqia, Delmont 101, 130
Warne, Russ J. 81
Warren, Sean 81
Washington, Perry J. 38, 81
Watson, Emmanuel 81
Weatherly, Ronald D. 81, 133
Weber, Haslip (Marty) 133
Weir, John M. 81
Wells, George B. 81
West, Scott 81
White, Darin M. 81
White, Darrin 42
White, Lloyd 111
White, Ralph E. 81
White, Raymond A. 42, 81
Whitehead, Elliott 19, 21, 39
Whitmarsh, Charles 81
Whitty, James D. 81
Williams, Greg 34, 81
Williams, James 81
Williams, Jolene 82
Williams, Kevin 82
Williams, Ulysses F. 82
Wittmer, William C. 82, 129
Wojtkiewicz, Joshua 82
Wong, Frank 112
Wong, Linda 82
Woodard, Terry 82
Worthington, Michael 82
Wraa, Alan N. 82, 95, 123, 124, 125, 128, 131
Wright, Harriet 82
Wright, Ricky 82
Wu, Eva 82
Wylie, Dennis-Ray 82

— Y —
Ybarra, David 82, 110
Yellin, Eya 83
Yi, David 82
Young, Lawrence 82, 110, 122
Young, Rick 36, 119
Young, Robert S. 83

— Z —
Zeisse, Rodney B. 83, 118